The Theory of
LINEAR
VISCOELASTICITY

The Theory of
LINEAR
VISCOELASTICITY

D. R. Bland
University of Manchester

DOVER PUBLICATIONS, INC.
Mineola, New York

Bibliographical Note

This Dover edition, first published in 2016, is an unabridged republication of the work originally published as Vol. 10 in the "International Series of Monographs on Pure and Applied Mathematics" by Pergamon Press, New York, in 1960.

Library of Congress Cataloging-in-Publication Data

Names: Bland, D. R. (David Russell), author.
Title: The theory of linear viscoelasticity / D.R. Bland.
Other titles: Linear viscoelasticity
Description: Mineola, New York : Dover Publications, Inc., 2016. | "This
  Dover edition, first published in 2016, is an unabridged republication of
  the work originally published as Vol. 10 in the "International Series of
  Monographs on Pure and Applied Mathematics" by Pergamon Press, New
  York, in 1960"—Title page verso. | Includes bibliographical references and index.
Identifiers: LCCN 2016015242| ISBN 9780486462691 | ISBN 0486462692
Subjects: LCSH: Elasticity. | Viscosity.
Classification: LCC QA931 .B5 2016 | DDC 531/.382—dc23 LC
  record available at https://lccn.loc.gov/2016015242

Manufactured in the United States by RR Donnelley
46269201   2016
www.doverpublications.com

CONTENTS

PAGE

1. MODELS—AN INTRODUCTION TO THE CONCEPTS OF VISCOELASTICITY

1. Force-extension equations of the simple models . 1
2. Creep and relaxation behaviour 5
3. Complex modulus and compliance . . . 12
4. Stored and dissipated energies 14
5. Creep and relaxation behaviour of some real materials 14

2. THE FOUNDATIONS OF THREE-DIMENSIONAL LINEAR VISCOELASTICITY

1. Hypotheses 19
2. The mechanics of the microscopic network . . 22
3. Introduction of normal co-ordinates . . . 25
4. Separation into deviatoric and dilatational components 29
5. A lemma 34
6. Creep and relaxation functions 36
7. Sinusoidal oscillations; complex modulus and compliance 40
8. Operational form of the stress–strain equation . 44
9. Model representation 48
10. Retardation and relaxation spectra . . . 50
11. Summary of the results of Chapter 2 . . . 55

3. STRESS ANALYSIS I: SINUSOIDAL OSCILLATION PROBLEMS

1. Stress analysis in viscoelasticity 57
2. Propagation of sinusoidally oscillating waves in an infinite medium 58

CONTENTS

3. The correspondence principle for sinusoidal oscillations 67
4. The vibrating reed 68
5. Free radial vibrations of a solid sphere . . 71
6. Rayleigh waves 73

4. STRESS ANALYSIS II: QUASI-STATIC PROBLEMS

1. The correspondence principle 76
2. Expansion of a reinforced cylinder by internal pressure 78
3. Point force on a semi-infinite plane . . . 84
4. Moving point force on a semi-infinite plane . . 86
5. Another form of the correspondence principle . 87
6. Indentation of an incompressible semi-infinite plane by a smooth rigid sphere 89
7. Biot's stability problem 91

5. STRESS ANALYSIS III: DYNAMIC PROBLEMS

1. The correspondence principle . . . 95
2. The propagation of longitudinal waves along semi-infinite rods 97
3. Normal impact on the boundary of a spherical cavity in an infinite medium 105
4. Normal impact on a clamped circular plate . . 109

6. MODEL FITTING TO MEASURED VALUES OF COMPLEX MODULUS OR COMPLIANCE

1. Procedure 113
2. First example 118
3. Second example 121

AUTHOR INDEX 125

CHAPTER 1

MODELS—AN INTRODUCTION TO THE CONCEPTS OF VISCOELASTICITY

1. Force-extension equations of the simple models

Viscoelasticity, as its name implies, is a generalization of elasticity and viscosity. The ideal linear elastic element is the spring. When a tensile force is applied to it, the increase in distance between its two ends is proportional to the force. The

FIG. 1.1. The elastic element, the spring.

FIG. 1.2. The viscous element, the dashpot.

ideal linear viscous element is the dashpot. It has the property that, when a tensile force is applied to it, the sides move apart at a rate that is proportional to the force. A loose fitting piston in a liquid-filled cylinder, arranged so that liquid flows out around the sides of the piston when it moves slowly up the cylinder, is an example of such an element. The two elements are represented diagrammatically in Figs. 1.1 and 1.2.

If F is the force and a is the extension, i.e. the increase in distance between the ends of the spring in the loaded from the unloaded state, then

$$F = Ea; \qquad (1)$$

E is a constant, known as the modulus of the spring. For the dashpot, the relationship is

$$F = \eta Da \qquad (2)$$

where the operator D denotes differentiation with respect to time; η is a constant, known as the viscosity of the dashpot.

In viscoelasticity the basic elastic and viscous elements are combined. In this chapter we shall be concerned with networks that are essentially one-dimensional. The elements all lie in the same direction and all the forces concerned act in this direction. In this section some of the simpler combination of elements are considered.

The elements can be combined in series. The combination is known as a Maxwell element and is illustrated in Fig. 1.3. Suppose a tensile force F is applied to this element. What is the relationship between the force and the extension? F is the same in the elastic and viscous elements and a is the sum of the extensions in the elastic and viscous elements, i.e.

$$a = a_{AB} = a_{AC} + a_{CB} \tag{3}$$

where C is the point of junction of spring and dashpot. a_{AC} is given by eq. (1) and Da_{CB} by eq. (2). a_{AB} and a_{CB} are eliminated from eqs. (1) to (3). Differentiating eq. (3),

$$Da = Da_{AC} + Da_{CB}$$

$$Da = \frac{1}{E} DF + \frac{1}{\eta} F \tag{4}$$

on substituting from eqs. (1) and (2). Equation (4) is the required relationship between a and F.

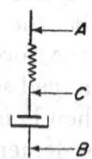

FIG. 1.3. The Maxwell element.

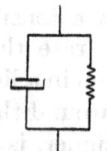

FIG. 1.4. The Voigt element.

The combination of the elastic and viscous elements in parallel is known as a Voigt element (sometimes as a Kelvin or Kelvin–Voigt element) and is illustrated in Fig. 1.4. The extension a is the same in spring and dashpot. If the tensile force in the spring is F_S and that in the dashpot F_D, then

$$F = F_S + F_D. \tag{5}$$

Substituting from eqs. (1) and (2),

$$F = Ea + \eta Da. \tag{6}$$

This is the relationship between force and extension for the Voigt element.

If E is infinite in eq. (4), eq. (4) is identical with eq. (2) and, if η is infinite, it is identical with eq. (1). If E is zero in eq. (6), eq. (6) is identical with eq. (2), and, if η is zero, it is identical with eq. (1). In these special cases the Maxwell and Voigt elements degenerate to the simplest elements.

From two springs and one dashpot, two non-degenerate[†] models[‡] can be made. They are illustrated in Fig. 1.5.

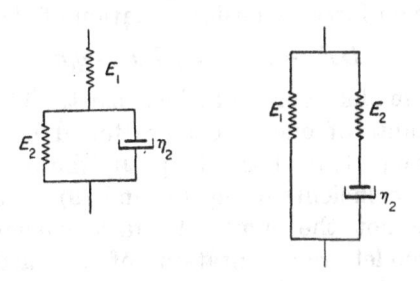

FIG. 1.5. Alternative representations of the three-element elastic model.

The left-hand model consists of an elastic and a Voigt element in series. The force F is the same in the two elements and the extension a is the sum of the extensions, a_1 and a_2, in the two elements, i.e.

$$a = a_1 + a_2.$$

From eq. (1), $\quad F = E_1 a_1$

and, from eq. (6), $\quad F = E_2 a_2 + \eta_2 D a_2.$

Eliminating a_1 and a_2,

$$DF + \frac{E_1 + E_2}{\eta_2} F = E_1 Da + \frac{E_1 E_2}{\eta_2} a. \tag{7}$$

† Two springs in series or in parallel can be replaced by a single spring as far as the mechanical behaviour of the combination is concerned. The words "non-degenerate" are intended to exclude this possibility. The same remarks apply to two dashpots in series or in parallel.

‡ The word "model" is used for the complete structure if it consists of more than two of the simplest elements.

Since an equation is unaltered when multiplied throughout by any non-zero quantity, the coefficient of the highest order derivative of F can be taken as unity.

The right-hand model in Fig. 1.5 consists of an elastic and a Maxwell element in parallel. It is left as an exercise for the reader to show that it has a force-extension relationship

$$DF + \frac{E_2'}{\eta_2'} F = (E_1' + E_2')Da + \frac{E_1'E_2'}{\eta_2'} a. \tag{8}$$

Thus for a given force-extension equation of the form

$$DF + p_0 F = q_1 Da + q_0 a \tag{9}$$

two distinct model representations exist. The elastic and viscous constants of one model are found by equating the coefficients of eq. (7) to those of eq. (9); those of the other by equating the coefficients of eq. (8) and (9). The two sets of constants are not the same. In fact whenever the non-degenerate model representation of a particular force-extension equation consists of more than two elements, the model representation is not unique.

The reader should now be able to show that two non-degenerate models can be constructed from two dashpots and one spring. Either model is known as the 3-element viscous model and its force-extension equation is of the form

$$DF + p_0 F = q_2 D^2 a + q_1 Da.$$

A particularly useful model is that shown in Fig. 1.6 with three more representations equivalent to it.

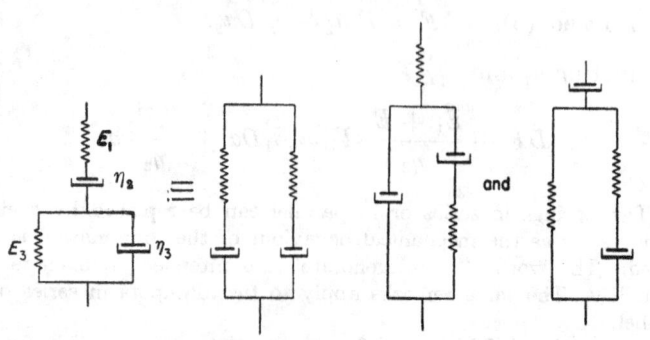

FIG. 1.6. The four element model of the first type and equivalent representations.

Using a self-evident notation

$$a = a_1 + a_2 + a_3.$$

From eq. (1) $\quad F = E_1 a_1,$

from eq. (2) $\quad F = \eta_2 D a_2$

and from eq. (6) $\quad F = E_3 a_3 + \eta_3 D a_3.$

Eliminating a_1, a_2 and a_3,

$$D^2 F + \left(\frac{E_1}{\eta_2} + \frac{E_1}{\eta_3} + \frac{E_3}{\eta_3}\right) DF + \frac{E_1 E_3}{\eta_2 \eta_3} F = E_1 D^2 a + \frac{E_1 E_3}{\eta_3} Da. \tag{10}$$

This is the force-extension relationship for the four element model of the first type.

2. Creep and relaxation behaviour

A creep test consists of applying a force to a model previously unloaded and then maintaining this force constant. The extension is measured as a function of the time. A relaxation test consists of imposing a constant extension and measuring the force required to do so as a function of the time. The response of the simple models to creep and relaxation will be considered.

The problem of creep can be formulated in mathematical terms thus: given $F = CH(t)$ and a relationship between F and a, find a. C is constant and $H(t)$ is the Heaviside unit function.

For the elastic element, the F–a relation is eq. (1) and so $a(t)$ is given by

$$a(t) = \frac{1}{E} F(t) = \frac{C}{E} H(t).$$

For the viscous element, eq. (2) gives

$$Da(t) = \frac{C}{\eta} H(t).$$

On integration

$$a(t) = \frac{C}{\eta} t H(t), \quad \text{since} \quad a = 0 \quad \text{at} \quad t = 0.$$

For the Maxwell element, eq. (4) gives

$$Da = \frac{C}{E}\delta(t) + \frac{C}{\eta}H(t)$$

where $\delta(t)$ is the Dirac δ-function.† On integration

$$a(t) = C\left(\frac{1}{E} + \frac{1}{\eta}t\right)H(t) \quad \text{since} \quad a = 0 \quad \text{at} \quad t = 0. \quad (11)$$

For the Voigt element, eq. (6) gives

$$Ea + \eta Da = CH(t).$$

Multiplying through by the integrating factor $\exp\{(E/\eta)t\}$,

$$\eta D\left\{\exp\left(\frac{E}{\eta}t\right)a\right\} = C\exp\left(\frac{E}{\eta}t\right)H(t).$$

On integration

$$a(t) = \frac{C}{E}\left\{1 - \exp\left(-\frac{E}{\eta}t\right)\right\}H(t) \quad \text{since} \quad a = 0 \quad \text{at} \quad t = 0. \quad (12)$$

The various extension responses are plotted in Fig. 1.7.

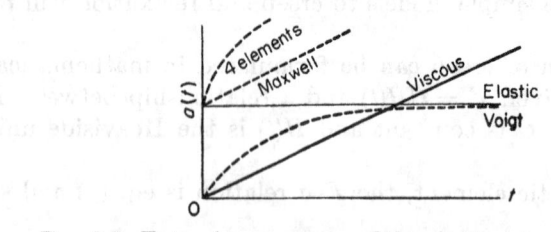

FIG. 1.7. Extension responses of the elastic, viscous, Maxwell, Voigt elements and four element model of the first type to unit force $H(t)$.

The extension response of the Maxwell element is the sum of the extension responses of the viscous and elastic elements. This is because it consists of these two elements in series—the stress is the same in each element and the overall extension is the sum of the extensions in the two elements. This property,

† $\delta(t) = 0$ for $t \neq 0$ and $\int_{-\epsilon}^{\epsilon}\delta(t)\,dt = 1$ for any $\epsilon > 0$. $\delta(t) = DH(t)$.

the extension response of elements combined in series, can be used to determine the extension response of more complicated models. For example, the four element model of Fig. 1.6 consists of a Maxwell and a Voigt element combined in series. Its extension response is therefore, using eqs. (11) and (12),

$$a(t) = C\left(\frac{1}{E_1} + \frac{1}{\eta_2} t\right) H(t) + \frac{C}{E_3} \left\{1 - \exp\left(-\frac{E_3}{\eta_3} t\right)\right\} H(t).$$

This response is plotted in Fig. 1.7.

Consider a model consisting of a large number of Voigt elements and a spring and a dashpot all connected in series. This is known as a generalized Voigt model and is illustrated in Fig. 1.8.

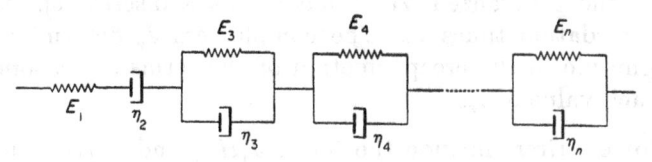

Fig. 1.8. The generalized Voigt model.

Its extension response will be

$$a(t) = C\left(\frac{1}{E_1} + \frac{1}{\eta_2} t\right) H(t) + C \sum_{r=3}^{n} \frac{1}{E_r} \left\{1 - \exp\left(-\frac{E_r}{\eta_r} t\right)\right\} H(t).$$

The extension response to unit force is sufficiently important to be given a special name. When the components corresponding to instantaneous elasticity and long term viscous flow are excluded, the response is known as "a creep function", denoted by $\psi(t)$. Hence for the generalized Voigt model

$$\psi(t) = \sum_{r=3}^{n} \frac{1}{E_r} \left\{1 - \exp\left(-\frac{E_r}{\eta_r} t\right)\right\} H(t). \qquad (13)$$

From eq. (12) it is seen that the time, τ, taken for the creep function of the Voigt element to reach a proportion $1 - (1/e)$ of its final value is

$$\tau = \frac{\eta}{E}.$$

τ is known as the retardation time of the Voigt element. The reciprocal of the modulus of the spring is known as the compliance and is denoted by J:

$$J = \frac{1}{E}.$$

Equation (12) can now be written

$$\psi(t) = J\{1 - \exp(-t/\tau)\}H(t). \tag{14}$$

Similarly if $\tau_r = \eta_r/E_r$ and $J_r = \dfrac{1}{E_r}$, eq. (13) can be written

$$\psi(t) = \sum_{r=3}^{n} J_r \{1 - \exp(-t/\tau_r)\}H(t). \tag{15}$$

Thus the generalized Voigt model has a discrete spectrum of retardation times τ_r. The compliances J_r determine the magnitude in the creep function of the terms corresponding to each value of τ_r.

Some writers include the terms $J_1 H(t)$ and $\dfrac{1}{\eta_2} t H(t)$ in the creep function $\psi(t)$. They correspond to $\tau_1 = 0$ and $\tau_2 = \infty$ respectively. The latter case arises because $E_2 = 0$. As $E_2 \to 0$, $\tau_2 \to \infty$ and

$$\frac{1}{E_2}\left\{1 - \exp\left(-\frac{E_2}{\eta_2}t\right)\right\} = \frac{1}{E_2}\left(\frac{E_2}{\eta_2}t - \frac{1}{2!}\left(\frac{E_2}{\eta_2}t\right)^2 + \ldots\right)$$

$$\to \frac{1}{\eta_2} t \quad \text{as} \quad E_2 \to 0.$$

A brief examination of the form of the creep function for a few particular cases in a given paper or book will show which of the alternative definitions of $\psi(t)$ the author has adopted.

If the number of Voigt elements in the generalized model increases indefinitely, in such a way that $J_r \to 0$ for all r but $\sum_{r=3}^{n} J_r \xrightarrow[n \to \infty]{}$ constant > 0, then the creep function becomes

$$\psi(t) = H(t) \int_0^\infty j(\tau)\{1 - \exp(-t/\tau)\}\,d\tau. \tag{16}$$

$j(\tau)$ is known as the distribution function of retardation times or more briefly the "retardation spectrum".

It is possible to visualize a model which has both a discrete and a continuous spectrum of relaxation times. For example, a new model could be made by placing one with a discrete spectrum and one with a continuous spectrum in series. In this case the creep function can be represented by a Stieltjes integral

$$\psi(t) = H(t) \int_0^\infty \{1 - \exp(-t/\tau)\} \, d\alpha(\tau), \qquad (17)$$

where $\alpha(\tau)$ is an increasing function of τ with derivatives at all but a finite number of points. At all points, except this finite number, the integrand is of the form of that in eq. (16); at the other points it contributes terms of the form of those in eq. (15).

Now consider relaxation tests. The problem of relaxation can be formulated in mathematical terms thus: given $a = KH(t)$ and a relationship between F and a, find F. K is constant. The force responses for the simple models will now be found.

For the elastic element, eq. (1),

$$F(t) = Ea(t) = KEH(t);$$

For the viscous element, eq. (2),

$$F(t) = \eta Da(t) = K\eta \, \delta(t).$$

For the Maxwell element, eq. (4),

$$F(t) = KE \exp\left(-\frac{E}{\eta} t\right) H(t) \qquad (18)$$

and for the Voigt element, eq. (6),

$$F(t) = KEH(t) + K\eta \, \delta(t). \qquad (19)$$

Two of the force responses just obtained contain a δ-function; the stress goes to infinity at the instant of application of the strain. The cause is that a dashpot, unlike a spring, cannot give a finite instantaneous extension response to a finite instantaneous change in force. Consequently, if a finite instantaneous extension is to be imposed on a dashpot, it requires an infinite force. For a model containing more than one element a finite instantaneous extension can be produced by a finite force only if there are one or more springs free to expand or contract without at the same time expanding or contracting a dashpot. For example the spring in the Maxwell element is free, that in the Voigt element is not. Consequently eq. (18) does not contain a δ-function, eq. (19) does. A model that can respond with an instantaneous finite change in extension to an instantaneous change in force is said to exhibit "instantaneous elasticity".

If reference is made back to the extension responses in creep, it will be seen that, as $t \to \infty$, $a(t)$ either tends to a limit or $a(t) \div Ct/\eta_2$ tends to unity. In the latter case a dashpot is free to flow unrestricted by any spring and the model is said to exhibit "long term viscous flow". The viscous element and the Maxwell element have this property, the elastic and the Voigt do not. In relaxation a dashpot, if one exists free from restriction by a spring, will eventually take up the whole extension and the force will drop to zero. The reader can check from the above equations for $F(t)$ that the force eventually falls to zero for the viscous and Maxwell elements but not for the elastic and Voigt.

Just as the constancy of the force enabled the extension responses in creep to be found by addition for models built up from simpler models in series so the constancy of the extension enables the force response in relaxation to be found by addition for models built up from simpler models in parallel. For example, the force response of the Voigt element is the sum of those for the elastic and for the viscous elements.

Consider a model consisting of a large number of Maxwell elements, a spring and a dashpot all in parallel, Fig. 1.9.

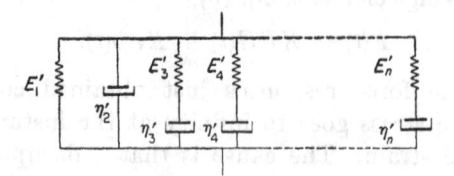

FIG. 1.9. The generalized Maxwell model.

The force response in relaxation of such a model will be

$$F(t) = KE_1'H(t) + K\eta_2'\,\delta(t) + K\sum_{r=3}^{n} E_r' \exp\left(-\frac{E_r'}{\eta_r'}t\right)H(t).$$

The force response to unit extension, $a(t) = H(t)$, but excluding the constant and δ-function components, is known as "the relaxation function", denoted by $\chi(t)$. Hence for the generalized Maxwell model

$$\chi(t) = \sum_{r=3}^{n} E_r' \exp\left(-\frac{E_r'}{\eta_r'} t\right). \tag{20}$$

If relaxation times τ_r' are defined by
$$\tau_r' = \eta_r'/E_r',$$
then
$$\chi(t) = \sum_{r=3}^{n} E_r' \exp(-t/\tau_r') H(t). \tag{21}$$

Thus the generalized Maxwell model has a discrete spectrum of relaxation times τ_r'. The elasticities E_r' determine the magnitude in the relaxation function of the terms corresponding to each value of τ_r'.

Some authors include the terms $E_1' H(t) + \eta_2' \delta(t)$ in the relaxation function. The term $E_1' H(t)$ corresponds to $\tau_1' = \infty$, i.e. $\eta_1' = \infty$; and the term† $\eta_2' \delta(t)$ corresponds to $\tau_2' = 0$, i.e. $E_2' = \infty$.

If the number of Maxwell elements in the generalized model increase indefinitely, in such a way that $E_r' \to 0$ but $\sum_{r=3}^{n} E_r' \xrightarrow[n \to \infty]{}$ constant > 0, then the relaxation function becomes

$$\chi(t) = H(t) \int_0^\infty y(\tau) \exp(-t/\tau) \, d\tau. \tag{22}$$

† Proof: For all values of t other than $t = 0$,
$$\lim_{E \to \infty} E \exp\left(-\frac{Et}{\eta}\right) H(t) = 0.$$

$$\int_{-\epsilon}^{\epsilon} E \exp\left(-\frac{Et}{\eta}\right) H(t) \, dt = \left[-\eta \exp\left(-\frac{Et}{\eta}\right) H(t)\right]_{-\epsilon}^{\epsilon}$$
$$+ \int_{-\epsilon}^{\epsilon} \eta \exp\left(-\frac{Et}{\eta}\right) \delta(t) \, dt \qquad \text{on integration by parts}$$
$$= -\eta \exp\left(-\frac{E\epsilon}{\eta}\right) + \eta \quad \therefore \lim_{E \to \infty} \int_{-\epsilon}^{\epsilon} E \exp\left(-\frac{Et}{\eta}\right) H(t) \, dt = \eta.$$

Therefore, by the definition of the δ-function,
$$\lim_{E \to \infty} E \exp\left(-\frac{Et}{\eta}\right) H(t) = \eta \delta(t).$$

$y(\tau)$ is known as the distribution function of relaxation times or more briefly the "relaxation spectrum". A model which has both a discrete and a continuous spectrum of relaxation times has a relaxation function represented by a Stieltjes integral

$$\chi(t) = H(t) \int_0^\infty \exp(-t/\tau) \, d\beta(\tau) \tag{23}$$

where $\beta(\tau)$ is an increasing function of τ with derivatives at all but a finite number of points.

3. Complex modulus and compliance

Let the model be subject to a sinusoidally oscillating force of radian frequency ω. After sufficient time has elapsed for the effect of the initial conditions to be negligible, the extension also will be of radian frequency ω. If $R[z]$ denotes the real part of z, then for sinusoidal oscillations

$$F = R[F^0(\omega) \exp(i\omega t)] \tag{24}$$

and

$$a = R[a^0(\omega) \exp(i\omega t)], \tag{25}$$

where $F^0(\omega)$ and $a^0(\omega)$ are independent of t and in general complex.

When the model is a spring, substitute, from eqs. (24) and (25) into eq. (1):

$$R[F^0(\omega) \exp(i\omega t)] = E R[a^0(\omega) \exp(i\omega t)].$$

For this equation to be valid for all t,

$$\frac{F^0(\omega)}{a^0(\omega)} = E.$$

In this case the ratio $\dfrac{F^0(\omega)}{a^0(\omega)}$ is equal to the modulus of the spring. When the model is a dashpot, from eqs. (24), (25) and (2)

$$R[F^0(\omega) \exp(i\omega t)] = \eta R[a^0(\omega) \, i\omega \exp(i\omega t)],$$

and therefore $\dfrac{F^0(\omega)}{a^0(\omega)} = i\omega\eta$.

Similarly for the Maxwell and Voigt models,

$$\frac{F^0(\omega)}{a^0(\omega)} = \left(\frac{1}{E} + \frac{1}{i\omega\eta}\right)^{-1} \text{ and } E + i\omega\eta \text{ respectively.}$$

The ratio $\frac{F^0(\omega)}{a^0(\omega)}$ for a particular model is known as "the complex modulus", $Y(i\omega)$ of that model. From eqs. (24) and (25) it is seen that the ratio has magnitude equal to the ratio of the amplitude of stress to that of strain and argument equal to the phase lag of the strain behind the stress. The reciprocal of the complex modulus is known as "the complex compliance", $J(i\omega)$. Both the complex modulus and the complex compliance are functions of the radian frequency ω.

If two models M' and M'' are placed in series, the force in each model is the same and the extension of the combination is the sum of the separate extensions. The complex compliances of the two models are denoted by $J'(i\omega)$ and $J''(i\omega)$ respectively and that of the combination by $J(i\omega)$. Then

$$J(i\omega) = \frac{a^0}{F^0} = \frac{a^{0\prime} + a^{0\prime\prime}}{F^0} = J'(i\omega) + J''(i\omega). \qquad (26)$$

This result is easily extended to show that the complex compliance of a number of models in series is the sum of their complex compliances. In particular the complex compliance of the generalized Voigt model, Fig. 8, is

$$J(i\omega) = \frac{1}{E_1} + \frac{1}{i\omega\eta_2} + \sum_{r=3}^{n} \frac{1}{E_r + i\omega\eta_r}. \qquad (27)$$

If two models M' and M'' are placed in parallel, the extension of each model is the same and the force in the combination is the sum of the separate forces. Denoting the complex moduli of the models and of the combination by $Y'(i\omega)$, $Y''(i\omega)$ and $Y(i\omega)$ respectively,

$$Y(i\omega) = \frac{F^0}{a^0} = \frac{F^{0\prime} + F^{0\prime\prime}}{a^0} = Y'(i\omega) + Y''(i\omega). \qquad (28)$$

Hence the complex modulus of a number of models in parallel

is the sum of their complex moduli. In particular the complex modulus of the generalized Maxwell model, Fig. 9, is

$$Y(i\omega) = E_1' + i\omega\eta_2' + \sum_{r=3}^{n} \left(\frac{1}{E_r'} + \frac{1}{i\omega\eta_r'}\right)^{-1}. \quad (29)$$

4. Stored and dissipated energies

Energy is stored in springs as elastic strain energy and energy is dissipated in dashpots as heat. No energy is dissipated in a spring and no energy stored in a dashpot. Consequently the total energy stored in a model, V, is equal to the sum of the energies stored in its component springs and the total rate of dissipation in a model, ϕ, is equal to the sum of the rates of dissipation in its component dashpots. Since the energy stored in a spring is equal to

$$\tfrac{1}{2}Ea^2 = \frac{1}{2E}F^2 = \tfrac{1}{2}Fa,$$

and the rate of dissipation in a dashpot is equal to $\eta(Da)^2 = (1/\eta)F^2 = FDa$,

$$V = \tfrac{1}{2}\sum_i E_i a_i^2 = \tfrac{1}{2}\sum_i F_i^2/E_i = \tfrac{1}{2}\sum_i F_i a_i \quad (30)$$

and $$\phi = \sum_i \eta_i (Da_i)^2 = \sum_i \frac{1}{\eta_i} F_i^2 = \sum_i F_i Da_i, \quad (31)$$

where the sums in V and ϕ are taken over all springs and over all dashpots in the model respectively.

5. Creep and relaxation behaviour of some real materials

In a creep test on a real material, external forces giving a simple distribution of stress in the material are applied at a particular time, held constant for a long period and then released. The displacements are measured at various times. From the known forces and displacements, the stress and strain components are calculated. Typical results† are shown in Figs. 1.10 [1] and 1.11 [2]. Typical results for relaxation tests are shown in Fig. 1.12 [3].

† Numbers in square brackets refer to the references given at the end of each chapter.

It can be seen from Fig. 1.10 that vulcanized rubber at 50 °C, 0 °C and −70 °C acts like a purely elastic material but at −40 °C and −50 °C it exhibits first instantaneous elasticity and then the strain continues to increase. At

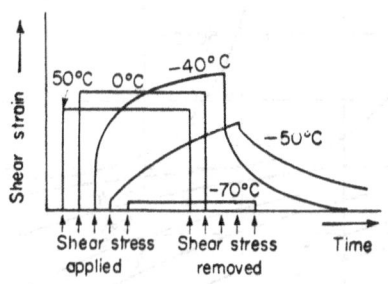

Fig. 1.10. Creep and recovery in shear of vulcanized rubber at different fixed temperatures.

−40 °C the strain tends to a limiting value; the test was not continued long enough to tell whether or not a limit is reached at −50 °C. The behaviour at −40 °C can be represented by a generalized Voigt model but without the dashpot in series.

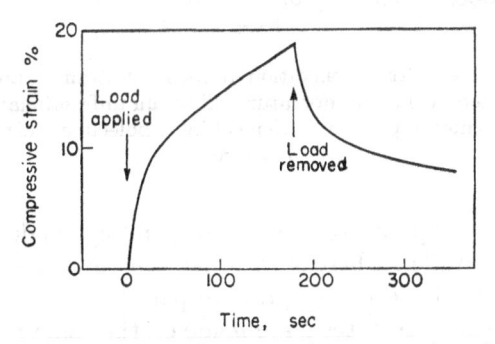

Fig. 1.11. Creep and recovery in compression of a nitrocellulose compound at 90 °C.

Similarly from Fig. 1.11 we deduce that the creep behaviour of nitrocellulose can be represented by a generalized Voigt model but without the spring in series.

The stress relaxation measurements on polyisobutylene, Fig. 1.12, can be represented by generalized Maxwell models. Since the stress falls to zero for large values of the time at

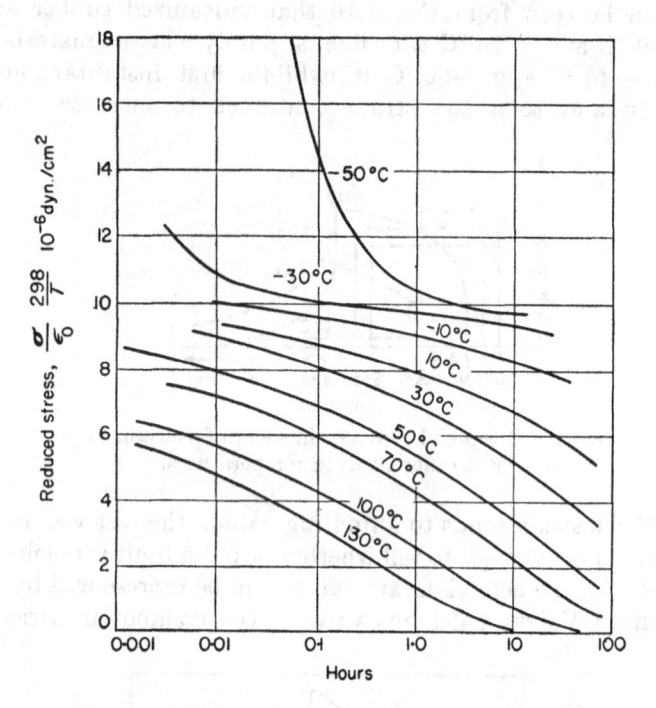

FIG. 1.12. Stress relaxation at constant strain ϵ_0 and at different fixed temperatures T for an unfractionated polymer of polyisobutylene of high molecular weight, σ = stress.

the higher temperatures the corresponding models have no spring in parallel, whereas at lower temperatures it appears that the models need the spring in parallel.

Suppose two creep tests are made on the same material but that the constant stress is different in the two cases. Then it is found experimentally that the strains, at the same time after application of stress, are proportional to the corresponding stresses provided that neither stress is too large. Linear theory is only exact for infinitesimal strains but it is a good approximation provided that the stress is not too large. The limits of validity are different for different materials and differ for the same material at different temperatures.

In an attempt to understand results, like those shown in

Figs. 1.10 to 1.12, many workers have made hypotheses about the microscopic structure of particular materials. A review of this work has been given by Ferry [4]. These microscopic structures have an important feature in common, namely that they are all mechanically equivalent to a network of springs and dashpots. In the next chapter the theory of three-dimensional viscoelasticity is developed from the hypothesis that the microscopic structure of the material is mechanically equivalent to a network of springs and dashpots, without any restriction on the number of these elements or on their arrangement. Linearity is ensured by postulating that the relative rotations of the elements are small. Temperature effects are not included in the theory. It is assumed that variations in temperature in a particular problem are small enough to neglect any change due to temperature variation in the mechanical properties of the material. If a viscoelastic material is deformed on separate occasions at different temperatures, then the mechanical properties are generally different on each occasion.

Other approaches to viscoelasticity have been adopted by different authors. The phenomenological approach, used by Gross [5] and by Staverman and Schwarzl [6] postulates the existence of creep functions and applies the Boltzmann superposition principle. This approach is subject to the objection that it does not seem possible to use it to determine the stored and dissipated energies. The assumption that the one-dimensional stress-strain law deducible from models can be generalized to three dimensions has been made by many writers, including Alfrey [7] and the writer of this book [8]. This procedure is subject to the objection that the microscopic structure of the material is three-dimensional; one cannot assume a priori that the mechanical properties of a three-dimensional network exhibit no features not shown in the one-dimensional case. Biot [9] uses the principles of irreversible thermodynamics. He assumes the existence of internal coordinates with which quadratic potential and dissipation functions can be associated. The present author feels that this approach is equivalent to that to be adopted in Chapter 2; thermodynamic postulates replace mechanical postulates and the results of the two approaches are identical.

REFERENCES

1. H. LEADERMAN: *Rheology*, Vol. 2, 6. Ed. F. R. Eirich, Academic Press, New York, 1956.
2. E. B. ATKINSON: *Rheology*, Vol. 2, 253. Ed. F. R. Eirich, Academic Press, New York, 1956.
3. R. D. ANDREWS and A. V. TOBOLSKY: *J. Polymer Sci.* **7** (1951) 221.
4. J. D. FERRY: *Physik der Hochpolymeren*, Vol. 4 (Chapter 1, Section 8, Springer, Berlin, 1956).
5. B. GROSS: *Mathematical Structure of the Theories of Viscoelasticity* (Herman, Paris, 1953).
6. A. J. STAVERMAN and L. SCHWARZL: *Physik der Hochpolymeren*, Vol. 4 (Chapter 1, Sections 1-7, Springer, Berlin, 1956).
7. T. ALFREY: *Mechanical Behaviour of High Polymers* (Interscience, New York, 1948).
8. D. R. BLAND: Proceedings of the conference on the properties of materials at high rates of strain, Institution of Mechanical Engineers, 1957, p. 156.
9. M. A. BIOT: *J. Appl. Phys.* 9, **25** (1954) 1385.

CHAPTER 2

THE FOUNDATIONS OF THREE-DIMENSIONAL LINEAR VISCOELASTICITY†

1. Hypotheses

Hypothesis 1

The microscopic structure of a linear viscoelastic material is mechanically equivalent to a network of linear viscous and elastic elements.

It was pointed out in the last section of Chapter 1 that this hypothesis is consistent with the microscopic models for particular materials which have been proposed to explain particular experimental results. For conciseness a linear elastic element is henceforth referred to as a "spring", a linear viscous element as a "dashpot".

Lamb commences his treatise on hydrodynamics with the words: "The following investigations proceed on the assumption that the matter with which we deal may be treated as practically continuous and homogeneous in structure, i.e. we assume that the properties of the smallest portions into which we can conceive it to be divided are the same as those of the substance in bulk". The same assumption is made in this book in treating macroscopic effects. However, the "smallest portion" of a viscoelastic material which satisfies this assumption is larger than the "smallest portion" of the liquids which Lamb was considering. It is consistent with the assumption of macroscopic homogeneity to assume that the microscopic structure of the material repeats itself at regular intervals, which are very small on the macroscopic scale. Hence:

Hypothesis 2

"The space occupied by a homogeneous viscoelastic material under no external forces can be subdivided into rectangular

† This chapter is taken from a paper by the author [1].

parallelepipeds of equal dimensions, close-packed and with their centres lying at the nodes of a rectangular lattice, such that the microscopic structure within and passing through the surface of each parallelepiped is identical". For conciseness a "rectangular parallelepiped" is henceforth referred to as a "box".

Even if each element is linear, the force-extension relationship of the boxes, into which the material has been divided, will only be linear if the relative rotations of neighbouring elements are small. This requires:

Hypothesis 3

"The extension in each element is small". The theory in this book will be limited to such extensions.

The question now arises of how to account for the fact that certain materials, normally classified as viscoelastic, are capable of exhibiting a steady rate of flow under constant stress and so, after a period of time, showing a large deformation. This effect is observed only for non cross-linked polymers, not for cross-linked polymers. The molecules of a non cross-linked polymer can slide over one another. After such a deformation the macroscopic shape of the polymer is altered but its properties are unaltered. It appears that the microscopic structure of the polymer is unaltered except that the component molecules have been rearranged. Hence:

Hypothesis 4

"To the strain in the macroscopic stress-strain law, to be found by considering the mechanical properties of the network within each box, assuming the relative positions of the molecules and hence of the boxes to be unaltered, can be added the strain produced by the relative motion of the molecules".

Since long term flow obeys the Newtonian viscous law under different types of stress and since no change of volume occurs, the deviatoric strain produced, $\bar{e}_{ij}$, is given by

$$\eta_{ijkl} D\bar{e}_{kl} = s_{ij},$$

where D denotes differentiation with respect to time, s_{ij} is the deviatoric stress tensor and η_{ijkl} is a constant tensor determined by the mechanical properties of the material. For an isotropic material η_{ijkl} must be isotropic. Therefore, for an isotropic material,

$$\eta D \bar{e}_{ij} = s_{ij}. \tag{1}$$

For a cross-linked polymer $(1/\eta) = 0$. It is convenient to refer to a viscoelastic material for which $(1/\eta) \neq 0$ as a fluid, to one for which $(1/\eta) = 0$ as a solid.

The type of deformation mechanism envisaged here for a viscoelastic fluid is similar to that shown by Bragg's bubble model [2]. In this model each bubble is able to deform slightly and layers of bubbles move over one another longitudinally and possibly diagonally—see Figs. 6 and 8 of the paper cited.

Unless the spacial stress derivatives on a macroscopic scale are very large, the change of stress over a distance equal to a few times the length of the side of a box will be very small and the overall deformation over a volume of which this distance is a typical length, will be effectively homogeneous; hence:

Hypothesis 5

"If A and B are any pair of corresponding points with co-ordinates $x_i^{(A)}$ and $x_i^{(B)}$ in the networks of adjacent boxes, then the displacements at A and B, $u_i^{(A)}$ and $u_i^{(B)}$ are related by

$$u_i^{(B)} = u_i^{(A)} + (\epsilon_{ij} + \xi_{ij})(x_j^{(B)} - x_j^{(A)}), \tag{2}$$

where the tensors ϵ_{ij} and ξ_{ij}, effectively constant in the regions of these boxes, are the macroscopic strain and rotation tensors respectively".

The strain $\bar{\epsilon}_{ij}$ due to relative molecular motion is not included in the strain ϵ_{ij} until notice is given to the contrary. When the spacial stress derivatives are discontinuous, the change in stress across a box will be large; but in actual examples the ratio of the number of boxes, over which the change in stress is large, to those in which it is nearly constant is likely to be so near to zero that the error, introduced by using Hypothesis 5 everywhere, will be negligible.

Since the properties of viscoelastic materials are being considered solely in the linear range, particle velocities will be small and large amounts of heat will not be generated in any small region by viscous dissipation. Hence:

Hypothesis 6

"The energy dissipated as heat is everywhere sufficiently small per unit volume to enable any variation in mechanical properties due to temperature change to be neglected". This hypothesis is generally made in the treatment of low speed viscous flow.

Lastly it is assumed that it is only necessary to consider inertial and body forces in the formulation of the equation of motion of each box, i.e.

Hypothesis 7

"In considering the equation of motion of any join,† inertial and body forces are small enough to be neglected".

2. The mechanics of the microscopic network

Until Section (10) it is assumed that each box contains a finite number of elements. These elements are joined at either end to one or more elements. The points of juncture are referred to as the "joins". Since there are only a finite number of joins and elements, the boxes can always be chosen so that no join lies in a box face and no element passes through the edge of a box. In these circumstances no element need be considered to lie in more than two boxes, which must have a common face, because if an element lies in more than two boxes it can be subdivided into smaller elements in series, each of which lies in only two adjacent boxes. Consider any one box. The mechanics of the network of elements will not be effected by a rotation of the material as a whole. Therefore the rotation of the material can always be chosen so that $\xi_{ij} = 0$ for the chosen box.

Let the joins inside the box be numbered 1 to n. Let the direction cosines of the element joining the rth join to the sth join be $l_i^{(rs)}$ in the undeformed state. Then the extension of this element when subject to stress is $l_i^{(rs)}(u_i^{(s)} - u_i^{(r)})$ plus a term of lower order. The relative rotation of elements has

† A "join" is defined in the next paragraph.

been restricted to be small and, as $\xi_{ij} = 0$ for the box considered, the absolute rotation of elements in the box is small.

If the element is a spring with elastic constant $E^{(rs)}$ the tension in the element is

$$E^{(rs)}l_j^{(rs)}(u_j^{(s)} - u_j^{(r)}).$$

Again neglecting a term of lower order, the tensile force in the element acts in the original direction of the element and is given by

$$E^{(rs)}l_i^{(rs)}l_j^{(rs)}(u_j^{(s)} - u_j^{(r)}).$$

Similarly, if the rth and sth joins are connected by a dashpot with viscous constant $\eta^{(rs)}$, the tensile force in the element is

$$\eta^{(rs)}l_i^{(rs)}l_j^{(rs)}(Du_j^{(s)} - Du_j^{(r)}).$$

Since inertial forces are neglected in viscoelasticity, except in formulating the macroscopic equations of motion (Hypothesis 7), the equation of motion of the rth join is

$$\sum_s E^{(rs)}l_i^{(rs)}l_j^{(rs)}(u_j^{(s)} - u_j^{(r)})$$
$$+ \sum_s \eta^{(rs)}l_i^{(rs)}l_j^{(rs)}(Du_j^{(s)} - Du_j^{(r)}) = 0, \qquad (3)$$

where the first sum is taken over all joins connected to r by a spring, and the second over all joins connected to r by a dashpot.

Equation (3) is valid for all joins r, $r = 1$ to n, lying inside the box. In general all the joins s connected to r do not lie in the box considered but all lie either in this box or in one of the six boxes which have a face in common with it. Let the edges of the box have lengths $L^{(k)}$, $k = 1, 2, 3$, and let $a_i^{(k)}$, $k = 1, 2, 3$, be unit vectors parallel to the edges of the box. If the join s lies in the box, entered by proceeding from any point in the original box in the positive direction of $a_i^{(k)}$, then by eq. (2) with $\xi_{ij} = 0$,

$$u_i^{(s)} = u_i^{(s')} + \epsilon_{ij}L^{(k)}a_j^{(k)}$$

where s' is the join in the original box corresponding to s. If the displacements at all points outside the box are replaced by the displacements at the corresponding points inside using the above equation, eq. (3) becomes

$$\sum_s E^{(rs)} l_i^{(rs)} l_j^{(rs)} (u_j^{(s')} + \epsilon_{jm} \sum_k L^{(k)} K^{(k)} a_m^{(k)} - u_j^{(r)})$$
$$+ \sum_s \eta^{(rs)} l_i^{(rs)} l_j^{(rs)} (Du_j^{(s')} + D\epsilon_{jm} \sum_k L^{(k)} K^{(k)} a_m^{(k)} - Du_j^{(r)}) = 0. \quad (4)$$

All the $K^{(k)}$ are zero unless either s lies in the box adjoining the original box in the positive direction of $a_i^{(k)}$, in which case the corresponding $K^{(k)}$ is $+1$ and the other two K zero, or s lies in the box adjoining the original box in the negative direction of $a_i^{(k)}$, in which case the corresponding $K^{(k)}$ is -1 and the other two K are zero.

The total force acting across the face of a box is equal to the sum of the forces in the elements crossing that face. It can be expressed in terms of the macroscopic stress tensor as $\pm \dfrac{T}{L^{(k)}} \sigma_{ij} a_j^{(k)}$ where T is the volume of a box.

Hence $\dfrac{T}{L^{(k)}} \sigma_{ij} a_j^{(k)} = \Sigma E^{(rs)} l_i^{(rs)} l_j^{(rs)} (u_j^{(s')} + \epsilon_{jm} L^{(k)} a_m^{(k)} - u_j^{(r)})$
$$+ \Sigma \eta^{(rs)} l_i^{(rs)} l_j^{(rs)} (Du_j^{(s')} + D\epsilon_{jm} L^{(k)} a_m^{(k)} - Du_j^{(r)}), \quad (5)$$

where the sums are taken over all elements crossing the face of the box whose outward normal is $+a_i^{(k)}$.

If V is the elastic energy stored in the springs per unit volume, then

$$2TV = \sum_{\substack{rs \\ r>s}} E^{(rs)} \{l_j^{(rs)} (u_j^{(s')} + \epsilon_{jm} \sum_k L^{(k)} K^{(k)} a_m^{(k)} - u_j^{(r)})\}^2 \quad (6)$$

and, if ϕ is the rate of dissipation of energy in the dashpots per unit volume,

$$T\phi = \sum_{\substack{rs \\ r>s}} \eta^{(rs)} \{l_j^{(rs)} (Du_j^{(s')} + D\epsilon_{jm} \sum_k L^{(k)} K^{(k)} a_m^{(k)} - Du_j^{(r)})\}^2 \quad (7)$$

where the sums are respectively over all springs and dashpots both inside the box and crossing the faces with outward normals $+a_i^{(k)}$.

Equations (4) and (5) are the equations of equilibrium, internal and external respectively, of the network inside a box. Equations (6) and (7) give the energy stored and dissipated within the box. Apart from the use of eqs. (4) to (7), no further reference to the microscopic structure of a viscoelastic material need be made in the subsequent analysis until Section (10).

3. Introduction of normal co-ordinates

We first observe that eqs. (4) and (5) are equivalent to

$$\frac{\partial V}{\partial u_j^{(r)}} + \frac{1}{2}\frac{\partial \phi}{\partial Du_j^{(r)}} = 0 \tag{8}$$

and

$$\frac{\partial V}{\partial \epsilon_{ij}} + \frac{1}{2}\frac{\partial \phi}{\partial D\epsilon_{ij}} = \sum_{k=1}^{3} \sigma_{il} a_l^{(k)} a_j^{(k)} \text{ respectively.} \tag{9}$$

V is a positive semi-definite quadratic form in the $3nu_j^{(r)}$, and in the nine ϵ_{ij}, and ϕ is a similar form in the $3nDu_j^{(r)}$ and in the nine $D\epsilon_{ij}$. Let ϕ' be the quadratic form identical to ϕ except that $Du_j^{(r)}$ is replaced by $u_j^{(r)}$ and $D\epsilon_{ij}$ by ϵ_{ij}. Let the ranks of V, ϕ and $2V + \phi'$ be R, S and M respectively. It will now be shown† that there exists a real linear transformation

$$u_j^{(r)} = \sum_{k=1}^{M} \gamma_{3r+j}^{(k)} \zeta_k \tag{10}$$

and

$$\epsilon_{ij} = \sum_{k=1}^{M} \gamma_{ij}^{(k)} \zeta_k, \tag{11}$$

such that

$$2V = \sum_{i=1}^{M-S} \zeta_i^2 + \sum_{i=M-S+1}^{R} (1 - \Lambda^{(i)}) \zeta_i^2 \tag{12}$$

and

$$\kappa^{-1}\phi = \sum_{i=M-S+1}^{R} \Lambda^{(i)} (D\zeta_i)^2 + \sum_{i=R+1}^{M} (D\zeta_i)^2, \tag{13}$$

where κ is a unit constant with the dimensions of time and $0 < \Lambda^{(i)} < 1$.

Since V is of the rank R, there exists a real linear transformation

$$u_j^{(r)} = \sum_{k=1}^{N} \alpha_{3r+j}^{(k)} \eta_k \quad \text{and} \quad \epsilon_{ij} = \sum_{k=1}^{N} \alpha_{ij}^{(k)} \eta_k$$

such that V is a positive definite form in $\eta_1, \eta_2 \ldots \eta_R$ and such that ϕ', though still positive semi-definite in $\eta_1, \eta_2 \ldots \eta_N$ is positive definite in $\eta_{R+1}, \eta_{R+2} \ldots \eta_N$ when $V = 0$. The rank of $2V + \phi'$ is still M after transformation, hence $N \geqslant M$. If $N > M$ there exist non-zero η_i for which $2V + \phi' = 0$.

† The author is indebted to Dr. W. Ledermann of Manchester University for this argument.

Since both V and ϕ' are non-negative, $2V + \phi' = 0$ implies $V = 0$ and $\phi' = 0$. But $V = 0$ implies $\eta_1, \eta_2 \ldots \eta_R = 0$ and $\phi' = 0$ implies $\eta_{R+1}, \eta_{R+2} \ldots \eta_N = 0$, which is a contradiction. Hence $N = M$.

Since $2V + \phi'$ is positive definite in the η_i, there exists a real linear transformation

$$\eta_j = \sum_{i=1}^{M} \beta_{ji} \zeta_i$$

such that
$$2V + \phi' = \sum_{i=1}^{M} \zeta_i^2$$

and
$$\phi' = \sum_{i=M-S+1}^{M} \Lambda^{(i)} \zeta_i^2.$$

Subtracting
$$2V = \sum_{i=1}^{M-S} \zeta_i^2 + \sum_{i=M-S+1}^{M} (1 - \Lambda^{(i)}) \zeta_i^2.$$

But V is of rank R and therefore $M - R$ of the $\Lambda^{(i)}$ must be unity. The ζ_i are ordered so that $\Lambda^{(i)} = 1$ for $i = R + 1$, $R + 2, \ldots M$. Since ϕ' and V are non-negative, $0 < \Lambda^{(i)} < 1$ for all $\Lambda^{(i)}$ in eqs. (12) and (13). The product of two real linear transformations is a real linear transformation. This proves the result stated in eqs. (10) to (13). The ζ_i are called "normal co-ordinates".

From eqs. (12) and (13),

$$\frac{\partial V}{\partial \zeta_i} + \frac{1}{2} \frac{\partial \phi}{\partial D\zeta_i} = (1 - \Lambda^{(i)}) \zeta_i + \kappa \Lambda^{(i)} D\zeta_i.$$

But
$$\frac{\partial V}{\partial \zeta_i} + \frac{1}{2} \frac{\partial \phi}{\partial D\zeta_i} = \sum_{r,j} \frac{\partial V}{\partial u_j^{(r)}} \frac{\partial u_j^{(r)}}{\partial \zeta_i} + \sum_{j,k} \frac{\partial V}{\partial \epsilon_{jk}} \frac{\partial \epsilon_{jk}}{\partial \zeta_i}$$
$$+ \frac{1}{2} \sum_{r,j} \frac{\partial \phi}{\partial Du_j^{(r)}} \frac{\partial Du_j^{(r)}}{\partial D\zeta_i} + \frac{1}{2} \sum_{j,k} \frac{\partial \phi}{\partial D\epsilon_{jk}} \frac{\partial D\epsilon_{jk}}{\partial D\zeta_i}$$
$$= \sum_{r,j} \gamma_{3r+j}^{(i)} \left(\frac{\partial V}{\partial u_j^{(r)}} + \frac{1}{2} \frac{\partial \phi}{\partial Du_j^{(r)}} \right)$$
$$+ \sum_{j,k} \gamma_{jk}^{(i)} \left(\frac{\partial V}{\partial \epsilon_{jk}} + \frac{1}{2} \frac{\partial \phi}{\partial D\epsilon_{jk}} \right) \text{ by eqs. (10) and (11)}$$
$$= \sum_{j,k} \gamma_{jk}^{(i)} \sum_{m=1}^{3} \sigma_{jl} a_l^{(m)} a_k^{(m)} \text{ by eqs. (8) and (9).}$$

Therefore
$$(1 - \Lambda^{(i)})\zeta_i + \kappa\Lambda^{(i)}D\zeta_i = \sum_{m=1}^{3} \gamma_{jk}^{(i)}\sigma_{jl}a_l^{(m)}a_k^{(m)}, \qquad (14)$$

where the summation sign over j and k has been omitted since summation is implied by the repeated suffices.

On integration for
$$M - S < i \leqslant R, \quad 0 < \Lambda^{(i)} < 1,$$
$$\zeta_i(t) = \int_{-\infty}^{t} \frac{1}{\kappa\Lambda^{(i)}} \sum_{m=1}^{3} \gamma_{jk}^{(i)} a_l^{(m)} a_k^{(m)} e^{-\lambda^{(i)}(t-\tau)} \sigma_{jl}(\tau)\, d\tau,$$

where
$$\lambda^{(i)} = \frac{1 - \Lambda^{(i)}}{\kappa\Lambda^{(i)}} \qquad (15)$$

and it is assumed that $\zeta_i \to 0$ as $t \to -\infty$.

On integration by parts
$$\zeta_i(t) = \frac{1}{\kappa\Lambda^{(i)}\lambda^{(i)}} \sum_{m=1}^{3} \gamma_{jk}^{(i)} a_l^{(m)} a_k^{(m)}$$
$$\times \int_{-\infty}^{t} [1 - \exp\{-\lambda^{(i)}(t-\tau)\}]\, d\sigma_{jl}(\tau). \qquad (16)$$

For $i \leqslant M - S$,
$$\zeta_i(t) = \sum_{m=1}^{3} \gamma_{jk}^{(i)} a_l^{(m)} a_k^{(m)} \sigma_{jl}(t), \qquad (17)$$

and for $i > R$,
$$\zeta_i(t) = \int_{-\infty}^{t} \frac{1}{\kappa} \sum_{m=1}^{3} \gamma_{jk}^{(i)} a_l^{(m)} a_k^{(m)} \sigma_{jl}(\tau)\, d\tau. \qquad (18)$$

To obtain a relation between stress and strain substitute for the ζ_i into eq. (11),

$$\epsilon_{ij} = \sum_{k=1}^{M} \gamma_{ij}^{(k)} \zeta_k = \sum_{k=1}^{M-S} \gamma_{ij}^{(k)} \sum_{m=1}^{3} \gamma_{pq}^{(k)} a_l^{(m)} a_q^{(m)} \sigma_{pl}(t)$$
$$+ \sum_{k=R+1}^{M} \gamma_{ij}^{(k)} \int_{-\infty}^{t} \frac{1}{\kappa} \sum_{m=1}^{3} \gamma_{pq}^{(k)} a_l^{(m)} a_q^{(m)} \sigma_{pl}(\tau)\, d\tau$$
$$+ \sum_{k=M-S+1}^{R} \gamma_{ij}^{(k)} \frac{1}{\kappa\Lambda^{(k)}\lambda^{(k)}} \sum_{m=1}^{3} \gamma_{pq}^{(k)} a_l^{(m)} a_q^{(m)}$$
$$\times \int_{-\infty}^{t} [1 - \exp\{-\lambda^{(k)}(t-\tau)\}]\, d\sigma_{pl}(\tau). \qquad (19)$$

Now the strains ϵ_{ij} given by eqs. (19) must be small for all finite values of σ_{ij}. If any terms appeared in the second sum in eqs. (19), $\epsilon_{ij} \to \infty$ as $t \to \infty$ for $\sigma_{ij} = H(t)A_{ij}$ where the A_{ij} are constant and non-zero. This is contrary to Hypothesis 3; therefore $R = M$ and

$$\epsilon_{ij}(t) = \sum_{k=1}^{R-S} \gamma_{ij}^{(k)} \sum_{m=1}^{3} \gamma_{pq}^{(k)} a_l^{(m)} a_q^{(m)} \sigma_{pl}(t)$$
$$+ \sum_{k=R-S+1}^{R} \gamma_{ij}^{(k)} \frac{1}{\kappa \Lambda^{(k)} \lambda^{(k)}} \sum_{m=1}^{3} \gamma_{pq}^{(k)} a_l^{(m)} a_q^{(m)}$$
$$\times \int_{-\infty}^{t} [1 - \exp\{-\lambda^{(k)}(t - \tau)\}] \, d\sigma_{pl}(\tau). \qquad (20)$$

The values of $\Lambda^{(i)}$, and hence of $\lambda^{(i)}$, are not in general distinct. Let the distinct values of $\Lambda^{(i)}$ be numbered from 1 to N. Then for a particular value of $\Lambda^{(i)}$, say $\Lambda^{(b)}$, the coefficient of $\int_{-\infty}^{t} [1 - \exp\{-\lambda^{(b)}(t - \tau)\}] \, d\sigma_{pl}(\tau)$ is

$$\frac{1}{\kappa \Lambda^{(b)} \lambda^{(b)}} \sum_{\Lambda^{(k)} = \Lambda^{(b)}} \gamma_{ij}^{(k)} \sum_{m=1}^{3} \gamma_{pq}^{(k)} a_l^{(m)} a_q^{(m)}.$$

This coefficient is multiplied by a second-order tensor σ_{ij} and contracted twice to form another second-order tensor ϵ_{ij}. Therefore the coefficient is a fourth-order tensor. But the coefficient is also a mechanical property of the material which has been assumed isotropic. Hence the coefficient must be an isotropic fourth-order tensor [3], i.e.

$$\frac{1}{\kappa \Lambda^{(b)} \lambda^{(b)}} \sum_{\Lambda^{(k)} = \Lambda^{(b)}} \gamma_{ij}^{(k)} \sum_{m=1}^{3} \gamma_{pq}^{(k)} a_l^{(m)} a_q^{(m)} = \frac{\alpha^{(b)}}{2} (\delta_{ip}\delta_{jl} + \delta_{il}\delta_{jp})$$
$$- \beta^{(b)} \delta_{ij}\delta_{pl} + \frac{\delta^{(b)}}{2} (\delta_{ip}\delta_{jl} - \delta_{il}\delta_{jp}). \qquad (21)$$

In general the sum of equal values of $\Lambda^{(k)}$ must contain at least six terms. This can be seen† by multiplying eq. (21) by $A_{ij}A_{pl}$ where A_{ij} is any second-order tensor:

$$\frac{1}{\kappa \Lambda^{(b)} \lambda^{(b)}} \sum_{\Lambda^{(k)} = \Lambda^{(b)}} \gamma_{ij}^{(k)} A_{ij} \sum_{m=1}^{3} \gamma_{pq}^{(k)} a_l^{(m)} a_q^{(m)} A_{pl}$$
$$= \frac{\alpha^{(b)} + \delta^{(b)}}{2} A_{ij}A_{ij} + \frac{\alpha^{(b)} - \delta^{(b)}}{2} A_{ij}A_{ji} - \beta^{(b)} A_{ii}A_{jj}.$$

† The author is indebted to Professor M. J. Lighthill of Manchester University for this argument.

The right-hand side, for general values of $\alpha^{(b)}$, $\beta^{(b)}$ and $\delta^{(b)}$ can be expressed in terms of three principal components of the symmetric part of the tensor A_{ij}, and the three components of the antisymmetric part and is therefore of rank 6. The left-hand side is the sum of squares because, if the x_i axis is parallel to $a_k^{(i)}$, $a_l^{(m)} = \delta_{lm}$ and $\sum_{m=1}^{3} \gamma_{pq}^{(k)} a_l^{(m)} a_q^{(m)} A_{pl} = \gamma_{pq}^{(k)} A_{pq}$, which is true for any choice of axes because both sides are invariant. Hence the left-hand side, to be of rank 6, must contain at least six terms. Since the left-hand side is non-negative for all A_{ij}, putting firstly $A_{ii} = 0$ with $A_{ij} = A_{ji}$, and secondly $A_{ij} = \delta_{ij}$

$$\alpha^{(b)} \geqslant 0 \quad \text{and} \quad \alpha^{(b)} - 3\beta^{(b)} \geqslant 0. \qquad (22)$$

The first sum on the right-hand side is treated similarly. Hence

$$\sum_{k=1}^{R-S} \gamma_{ij}^{(k)} \sum_{m=1}^{3} \gamma_{pq}^{(k)} a_l^{(m)} a_q^{(m)} = \frac{\alpha}{2}(\delta_{ip}\delta_{jl} + \delta_{il}\delta_{jp})$$
$$- \beta\delta_{ij}\delta_{pl} + \frac{\delta}{2}(\delta_{ip}\delta_{jl} - \delta_{ip}\delta_{jp}). \qquad (23)$$

Substitute from eqs. (21) and (23) into eqs. (20):

$$\epsilon_{ij}(t) = \alpha\sigma_{ij}(t) - \beta\delta_{ij}\sigma_{kk}(t) + \sum_{b=1}^{N} \int_{-\infty}^{t} (1 - \exp\{-\lambda^{(b)}(t-\tau)\})$$
$$\times (\alpha^{(b)} d\sigma_{ij}(\tau) - \beta^{(b)}\delta_{ij} d\sigma_{kk}(\tau)). \qquad (24)$$

Equation (24) is one form of the stress–strain relation for an isotropic viscoelastic solid. For an isotropic viscoelastic fluid, the strain $\bar{\epsilon}_{ij}$ produced by the relative molecular motion must be added to the right-hand side. Using eq. (1), this gives

$$\epsilon_{ij} = \alpha\sigma_{ij} - \beta\delta_{ij}\sigma_{kk} + \frac{1}{\eta}\int_{-\infty}^{t} s_{ij}(\tau)\, d\tau$$
$$+ \sum_{b=1}^{N} \int_{-\infty}^{t} [1 - \exp\{-\lambda^{(b)}(t-\tau)\}](\alpha^{(b)} d\sigma_{ij}(\tau)$$
$$- \beta^{(b)}\delta_{ij} d\sigma_{kk}(\tau)). \qquad (25)$$

4. Separation into deviatoric and dilatational components

If we introduce the deviatoric components of strain and stress, e_{ij} and s_{ij} respectively,

$$e_{ij} = \epsilon_{ij} - \tfrac{1}{3}\delta_{ij}\epsilon_{kk} \quad \text{and} \quad s_{ij} = \sigma_{ij} - \tfrac{1}{3}\delta_{ij}\sigma_{kk}, \qquad (26)$$

then from eq. (25)

$$e_{ij} = \alpha s_{ij} + \frac{1}{\eta} \int_{-\infty}^{t} s_{ij}(\tau) \, d\tau$$
$$+ \sum_{b=1}^{N} \int_{-\infty}^{t} \alpha^{(b)} [1 - \exp\{-\lambda^{(b)}(t - \tau)\}] \, ds_{ij}(\tau), \quad (27)$$

and

$$\epsilon_{kk} = \gamma \sigma_{kk} + \sum_{b=1}^{N} \int_{-\infty}^{t} \gamma^{(b)} [1 - \exp\{-\lambda^{(b)}(t - \tau)\}] \, d\sigma_{kk}(\tau), \quad (28)$$

where

$$\gamma^{(b)} = \alpha^{(b)} - 3\beta^{(b)} \quad \text{and} \quad \gamma = \alpha - 3\beta. \quad (29)$$

From eq. (22),

$$\alpha^{(b)} \geqslant 0 \quad \text{and} \quad \gamma^{(b)} \geqslant 0. \quad (30)$$

Conversely, eq. (25) can be derived from eqs. (27) and (28).

We shall now show that the stored energy per unit volume, V, can be expressed as sums of deviatoric and dilatational parts. Substituting from eq. (16), for $0 < \Lambda^{(i)} < 1$,

$$\sum_{\Lambda^{(i)} = \Lambda^{(b)}} (1 - \Lambda^{(i)}) \zeta_i^2 = \frac{1 - \Lambda^{(b)}}{(\kappa \Lambda^{(b)} \lambda^{(b)})^2} \sum_{\Lambda^{(i)} = \Lambda^{(b)}} \sum_{m=1}^{3} \gamma_{jk}^{(i)} a_l^{(m)} a_k^{(m)}$$
$$\times \sum_{n=1}^{3} \gamma_{pq}^{(i)} a_r^{(n)} a_q^{(n)} \int_{-\infty}^{t} \int_{-\infty}^{t} [1 - \exp\{-\lambda^{(b)}(t - \tau)\}]$$
$$\times [1 - \exp\{-\lambda^{(b)}(t - \theta)\}] \, d\sigma_{jl}(\tau) \, d\sigma_{pr}(\theta).$$

Using eqs. (15) and (21),

$$\sum_{\Lambda^{(i)} = \Lambda^{(b)}} (1 - \Lambda^{(i)}) \zeta_i^2 = \int_{-\infty}^{t} \int_{-\infty}^{t} [1 - \exp\{-\lambda^{(b)}(t - \tau)\}]$$
$$\times [1 - \exp\{-\lambda^{(b)}(t - \theta)\}] \sum_{m=1}^{3} \{\alpha^{(b)} a_l^{(m)} a_r^{(m)} \, d\sigma_{pl}(\tau) \, d\sigma_{pr}(\theta)$$
$$- \beta^{(b)} a_l^{(m)} a_k^{(m)} \, d\sigma_{kl}(\tau) \, d\sigma_{rr}(\theta)\}.$$

When the x_i axis is parallel to $a_k^{(i)}$, $a_k^{(i)} = \delta_{ik}$,

$$\sum_{m=1}^{3} a_l^{(m)} a_r^{(m)} \, d\sigma_{pl}(\tau) \, d\sigma_{pr}(\theta) = d\sigma_{pr}(\tau) \, d\sigma_{pr}(\theta)$$

and

$$\sum_{m=1}^{3} a_l^{(m)} a_k^{(m)} \, d\sigma_{kl}(\tau) = d\sigma_{kk}(\theta).$$

But both sides of these two equations are invariants and therefore true for any set of axes. Hence

$$\sum_{\Lambda^{(i)}=\Lambda^{(b)}} (1 - \Lambda^{(i)})\zeta_i^2 = \int_{-\infty}^t \int_{-\infty}^t [1 - \exp\{-\lambda^{(b)}(t - \tau)\}]$$
$$\times [1 - \exp\{-\lambda^{(b)}(t - \theta)\}](\alpha^{(b)} \, d\sigma_{jk}(\theta) \, d\sigma_{jk}(\tau)$$
$$- \beta^{(b)} \, d\sigma_{jj}(\theta) \, d\sigma_{kk}(\tau)).$$

If we introduce increments of deviatoric stress components

$$d\sigma_{jk}(\theta) \, d\sigma_{jk}(\tau) = ds_{jk}(\theta) \, ds_{jk}(\tau) + \tfrac{1}{3} \, d\sigma_{ii}(\theta) \, d\sigma_{jj}(\tau).$$

Therefore

$$\sum_{\Lambda^{(i)}=\Lambda^{(b)}} (1 - \Lambda^{(i)})\zeta_i^2 = \int_{-\infty}^t \int_{-\infty}^t [1 - \exp\{-\lambda^{(b)}(t - \tau)\}]$$
$$\times [1 - \exp\{-\lambda^{(b)}(t - \theta)\}](\alpha^{(b)} ds_{jk}(\theta) \, ds_{jk}(\tau)$$
$$+ \tfrac{1}{3}\gamma^{(b)} \, d\sigma_{jj}(\theta) \, d\sigma_{kk}(\tau)).$$

Similarly

$$\sum_{i=1}^{M-S} \zeta_i^2 = \alpha s_{jk} s_{jk} + \tfrac{1}{3}\gamma \sigma_{jj}\sigma_{kk}.$$

Substituting from the last two equations into eq. (12),

$$2V = \alpha s_{jk} s_{jk} + \sum_{b=1}^{N} \alpha^{(b)} \int_{-\infty}^t \int_{-\infty}^t [1 - \exp\{-\lambda^{(b)}(t - \tau)\}]$$
$$\times [1 - \exp\{-\lambda^{(b)}(t - \theta)\}] \, ds_{jk}(\theta) \, ds_{jk}(\tau)$$
$$+ \tfrac{1}{3}\gamma \sigma_{jj}\sigma_{kk} + \tfrac{1}{3}\sum_{b=1}^{N} \gamma^{(b)} \int_{-\infty}^t \int_{-\infty}^t [1 - \exp\{-\lambda^{(b)}(t - \tau)\}]$$
$$\times [1 - \exp\{-\lambda^{(b)}(t - \theta)\}] \, d\sigma_{jj}(\theta) \, d\sigma_{kk}(\tau). \quad (31)$$

The rate of dissipation of energy per unit volume, ϕ, can be treated in a similar manner. Equation (13) requires modification because $R = M$ and because it does not include the energy dissipated by the relative molecular motion—eq. (1). It becomes

$$\phi = \kappa \sum_{i=R-S+1}^{R} \Lambda^{(i)} (D\zeta_i)^2 + \frac{1}{\eta} s_{ij} s_{ij}. \quad (32)$$

The second term on the right-hand side vanishes for a viscoelastic solid because $(1/\eta) = 0$ for a solid. From eq. (16)

$$D\zeta_i(t) = \frac{1}{\kappa \Lambda^{(i)}} \sum_{m=1}^{3} \gamma_{jk}^{(i)} a_l^{(m)} a_k^{(m)} \int_{-\infty}^{t} \exp\{-\lambda^{(i)}(t-\tau)\} d\sigma_{jl}(\tau);$$

$$\kappa \sum_{\Lambda^{(i)}=\Lambda^{(b)}} \Lambda^{(i)}(D\zeta_i)^2 = \frac{1}{\kappa \Lambda^{(b)}} \sum_{\Lambda^{(i)}=\Lambda^{(b)}} \sum_{m=1}^{3} \gamma_{jk}^{(i)} a_l^{(m)} a_k^{(m)} \sum_{n=1}^{3} \gamma_{pq}^{(i)} a_r^{(n)} a_q^{(n)}$$

$$\times \int_{-\infty}^{t} \int_{-\infty}^{t} \exp\{-\lambda^{(i)}(t-\tau)\} \exp\{-\lambda^{(i)}(t-\theta)\} d\sigma_{jl}(\tau) d\sigma_{pr}(\theta).$$

Whence, as above,

$$\kappa \sum_{\Lambda^{(i)}=\Lambda^{(b)}} \Lambda^{(i)}(D\zeta_i)^2 = \lambda^{(b)} \int_{-\infty}^{t} \int_{-\infty}^{t} \exp\{-\lambda^{(b)}(2t-\tau-\theta)\}$$

$$\times (\alpha^{(b)} ds_{jk}(\theta) ds_{jk}(\tau) + \tfrac{1}{3}\gamma^{(b)} d\sigma_{jj}(\theta) d\sigma_{kk}(\tau)),$$

and

$$\phi = \frac{1}{\eta} s_{ij} s_{ij} + \sum_{b=1}^{N} \lambda^{(b)} \alpha^{(b)} \int_{-\infty}^{t} \int_{-\infty}^{t} \exp\{-\lambda^{(b)}(2t-\tau-\theta)\}$$

$$\times ds_{jk}(\theta) ds_{jk}(\tau)$$

$$+ \tfrac{1}{3} \sum_{b=1}^{N} \lambda^{(b)} \gamma^{(b)} \int_{-\infty}^{t} \int_{-\infty}^{t} \exp\{-\lambda^{(b)}(2t-\tau-\theta)\} d\sigma_{jj}(\theta) d\sigma_{kk}(\tau).$$

(33)

From eqs. (27), (28), (31) and (33) there follow the important results:

"In a linear isotropic viscoelastic material, (i) each deviatoric component of strain is related by the stress–strain law solely to the corresponding deviatoric component of stress and the relationship is the same for each component; (ii) the dilatational part of the strain is related solely to the dilatational part of the stress; and (iii) both the stored energy and the dissipation are the sum of ten terms, nine of these terms with six of them equal in pairs come from the nine deviatoric components, and are identical in form and the tenth comes from the dilatation. There are no product terms involving either any two deviatoric components or any one component and the dilatation".

It will therefore be unnecessary in the remainder of this

chapter to consider deviatoric and dilatational terms separately. We consider a strain–stress relationship

$$\epsilon = \frac{1}{E}\sigma + \frac{1}{\eta}\int_{-\infty}^{t}\sigma(\tau)\,d\tau$$
$$+ \sum_{b=1}^{N}\int_{-\infty}^{t} B^{(b)}[1 - \exp\{-\lambda^{(b)}(t-\tau)\}]\,d\sigma(\tau), \quad (34)$$

a stored energy per unit volume

$$2AV = \frac{1}{E}\sigma^2 + \sum_{b=1}^{N}\int_{-\infty}^{t}\int_{-\infty}^{t} B^{(b)}[1 - \exp\{-\lambda^{(b)}(t-\tau)\}]$$
$$\times [1 - \exp\{-\lambda^{(b)}(t-\theta)\}]\,d\sigma(\theta)\,d\sigma(\tau) \quad (35)$$

and a rate of dissipation per unit volume

$$A\phi = \frac{1}{\eta}\sigma^2 + \sum_{b=1}^{N}\int_{-\infty}^{t}\int_{-\infty}^{t} \lambda^{(b)} B^{(b)}$$
$$\times \exp\{-\lambda^{(b)}(2t - \tau - \theta)\}\,d\sigma(\theta)\,d\sigma(\tau). \quad (36)$$

The ϵ and σ can be interpreted as any deviatoric component or as the dilatation.

$$\left.\begin{array}{ll} \dfrac{1}{E} = \alpha,\ B^{(b)} = \alpha^{(b)},\ A = 1 & \text{for a deviatoric component} \\ \text{and}\ \ \dfrac{1}{E} = \gamma,\ B^{(b)} = \gamma^{(b)},\ A = 3 & \text{for the dilatation.} \end{array}\right\} \quad (37)$$

For a system involving more than one component, the stored energy and the dissipation are the sums of the respective quantities over all the components. Note that $(1/\eta) = 0$ for the dilatation. The microscopic model was constructed in such a way that the material always returns to its original volume after all external forces have been removed for a long period of time.

One-dimensional viscoelastic behavior has been considered by many authors. In such treatments all the microscopic elements are assumed to lie in the same direction and an equation of the form of eq. (34) can be derived [4]. For a rod, if the x_1 axis is taken along the rod and all stress components other than σ_{11} are put equal to zero, from eq. (25)

$$\epsilon_{11} = (\alpha - \beta)\sigma_{11} + \frac{2}{3\eta}\int_{-\infty}^{t} \sigma_{11}(\tau)\, d\tau$$

$$+ \sum_{b=1}^{N} (\alpha^{(b)} - \beta^{(b)})\,[1 - \exp\{-\lambda^{(b)}(t-\tau)\}]\, d\sigma_{11}(\tau), \quad (37a)$$

an equation of the form of eq. (34).

It can be verified from eqs. (25), (31) and (33) that the law of conservation of energy is satisfied, i.e.

$$DV + \phi = \sigma_{ij} D\epsilon_{ij}. \tag{38}$$

5. A lemma

In this section we prove a lemma and three corollaries.
Lemma: "If all θ_r and λ_r are positive and $\lambda_r < \lambda_{r+1}$, then

$$\left(\sum_{r=1}^{M} \frac{\theta_r}{p + \lambda_r}\right)^{-1} = E' + \eta' p + p \sum_{r=1}^{M-1} \frac{C_r}{p + \mu_r}$$

where E', η' all C_r and all μ_r are positive and $\lambda_r < \mu_r < \lambda_{r+1}$; the μ_r are ordered so that $\mu_{r+1} > \mu_r$.

Proof:

$$\sum_{r=1}^{M} \frac{\theta_r}{p + \lambda_r} = \left(\sum_{r=1}^{M} \theta_r\right) p_{M-1} \left(\prod_{r=1}^{M}(p + \lambda_r)\right)^{-1}$$

where p_{M-1} is a polynomial of degree $M-1$ in p with real positive coefficients and with the coefficient of p^{M-1} equal to 1. p_{M-1} can be expressed as

$$p_{M-1} = \prod_{r=1}^{M-1}(p + \mu_r)$$

where any of the μ_r, which are not real, are complex conjugate in pairs. Hence

$$\sum_{r=1}^{M} \frac{\theta_r}{p + \lambda_r} = \left(\sum_{r=1}^{M} \theta_r\right)\left(\prod_{r=1}^{M-1}(p + \mu_r)\right)\left(\prod_{r=1}^{M}(p + \lambda_r)\right)^{-1}.$$

By the theory of partial fractions

$$\theta_s = \left(\sum_{r=1}^{M} \theta_r\right)\left(\prod_{r=1}^{M-1}(-\lambda_s + \mu_r)\right)\left(\prod_{\substack{r=1 \\ r \neq s}}^{M}(-\lambda_s + \lambda_r)\right)^{-1}.$$

θ_s and $\sum_{r=1}^{M} \theta_r$ are positive. $\prod_{\substack{r=1 \\ r \neq s}}^{M}(-\lambda_s + \lambda_r)$ is positive for odd s, negative for even s. Therefore $\prod_{r=1}^{M-1}(-\lambda_s + \mu_r)$ is positive for odd s and negative for even s. If μ_r is complex, $(-\lambda_s + \mu_r)(-\lambda_s + \bar{\mu}_r)$ is positive for all λ_s where $\bar{\mu}_r$ denotes the complex conjugate of μ_r. If μ_r is real, $-\lambda_s + \mu_r$ changes sign once and only once as $-\lambda_s$ increases from $-\infty$ to $+\infty$. Now $\prod_{r=1}^{M-1}(-\lambda_s + \mu_r)$ changes sign $M-1$ times as λ_s takes the values $\lambda_1, \lambda_2 \ldots \lambda_M$ in turn. These changes are only consistent if one and only one factor in the product $\prod_{r=1}^{M-1}(-\lambda_s + \mu_r)$ changes sign whenever the suffix s increases by unity. This will occur only if all the μ_r are real and one and only one μ_r lies between successive λ_r. If the μ_r are ordered so that $\mu_{r+1} > \mu_r$, then we must have $\lambda_r < \mu_r < \lambda_{r+1}$. Since $\lambda_1 > 0$, $\mu_r > 0$. Now

$$\left(\sum_{r=1}^{M} \frac{\theta_r}{p + \lambda_r}\right)^{-1} = \left(\sum_{r=1}^{M} \theta_r\right)^{-1}\left(\prod_{r=1}^{M}(p + \lambda_r)\right)\left(\prod_{r=1}^{M-1}(p + \mu_r)\right)^{-1}.$$

Put
$$\eta' = \left(\sum_{r=1}^{M} \theta_r\right)^{-1}$$

and
$$E' = \left(\sum_{r=1}^{M} \theta_r\right)^{-1}\left(\prod_{r=1}^{M} \lambda_r\right)\left(\prod_{r=1}^{M-1} \mu_r\right)^{-1},$$

then $\left(\sum_{r=1}^{M} \dfrac{\theta_r}{p + \lambda_r}\right)^{-1} - E' - \eta' p$

$$= \left(\sum_{r=1}^{M} \theta_r\right)^{-1}\left(\prod_{r=1}^{M-1}(p + \mu_r)\right)^{-1}\left[\left(\prod_{r=1}^{M}(p + \lambda_r)\right)\right.$$
$$\left. - p\left(\prod_{r=1}^{M-1}(p + \mu_r)\right) - \left(\prod_{r=1}^{M} \lambda_r\right)\left(\prod_{r=1}^{M-1} \mu_r\right)^{-1}\prod_{r=1}^{M-1}(p + \mu_r)\right].$$

The term in the square brackets is of the form pP_{M-2}.

THE THEORY OF LINEAR VISCOELASTICITY

Therefore by partial fractions

$$\left(\sum_{r=1}^{M}\frac{\theta_r}{p+\lambda_r}\right)^{-1} - E' - \eta'p = p\sum_{r=1}^{M-1}\frac{C_r}{p+\mu_r}$$

where

$$C_s = \left(\sum_{r=1}^{M}\theta_r\right)^{-1}(-\mu_s)^{-1}\left(\prod_{r=1}^{M}(-\mu_s+\lambda_r)\right)\left(\prod_{\substack{r=1\\r\neq s}}^{M-1}(-\mu_s+\mu_r)\right)^{-1}.$$

Since $\lambda_r < \mu_r < \lambda_{r+1}$ and $\lambda_1 \geqslant 0$, $C_s > 0$.

This completes the proof of the lemma.

Corollary 1: If $\lambda_1 \to 0$, then $E' \to 0$.

Corollary 2: If $\lambda_M \to \infty$, $\theta_M \to \infty$ so that $\lambda_M \theta_M^{-1} \to E$,

then $\displaystyle\sum_{r=1}^{M}\frac{\theta_r}{p+\lambda_r} \to \frac{1}{E} + \sum_{r=1}^{M-1}\frac{\theta_r}{p+\lambda_r}$, $\eta' \to 0$,

$$E' \to E\left(\prod_{r=1}^{M-1}\lambda_r\right)\left(\prod_{r=1}^{M-1}\mu_r\right)^{-1},$$

$$C_s \to (-\mu_s)^{-1}\left(\prod_{r=1}^{M-1}(-\mu_s+\lambda_r)\right)\left(\prod_{\substack{r=1\\r\neq s}}^{M-1}(-\mu_s+\mu_r)\right)^{-1}.$$

Corollary 3: If $\lambda_1 \to 0$ and $\lambda_M \to \infty$, $\theta_M \to \infty$ so that $\lambda_M \theta_M^{-1} \to E$,

then $\displaystyle\sum_{r=1}^{M}\frac{\theta_r}{p+\lambda_r} \to \frac{1}{E} + \frac{\theta_1}{p} + \sum_{r=2}^{M-1}\frac{\theta_r}{p+\lambda_r}$, $\eta' \to 0$,

$$E' \to 0, C_s \to (-\mu_s)^{-1}\left(\prod_{r=1}^{M-1}(-\mu_s+\lambda_r)\right)\left(\prod_{\substack{r=1\\r\neq s}}^{M-1}(-\mu_s+\mu_r)\right)^{-1}.$$

6. Creep and relaxation functions

The creep function $\psi(t)$ is defined as the strain produced by a stress $H(t)$, the Heavyside unit function, but excluding the instantaneous response and the long term viscous flow. If we substitute $\sigma(t) = H(t)$ in eq. (34),

$$\epsilon(t) = \frac{1}{E}H(t) + \frac{1}{\eta}tH(t) + \sum_{b=1}^{N}B^{(b)}\{1 - \exp(-\lambda^{(b)}t)\}H(t).$$

Hence we have the result:

"The deviatoric and dilatational creep functions of a linear isotropic viscoelastic material whose microscopic structure contains only a finite number of elements are of the form

$$\psi(t) = \sum_{b=1}^{N} B^{(b)} \{1 - \exp(-\lambda^{(b)}t)\} H(t), \qquad (39)$$

where the $B^{(b)}$ and $\lambda^{(b)}$ are positive constants."

Substituting back from eq. (39) into eq. (34),

$$\epsilon(t) = \frac{1}{E}\sigma(t) + \frac{1}{\eta}\int_{-\infty}^{t} \sigma(\tau)\,d\tau + \int_{-\infty}^{t} \psi(t-\tau)\,d\sigma(\tau). \qquad (40)$$

The stored and dissipated energies can be expressed in terms of the stress at previous times and the creep function. From eq. (35)

$$2AV = \frac{1}{E}\sigma^2 + \sum_{b=1}^{N}\int_{-\infty}^{t}\int_{-\infty}^{t} B^{(b)}([1 - \exp\{-\lambda^{(b)}(t-\tau)\}]$$
$$+ [1 - \exp\{-\lambda^{(b)}(t-\theta)\}] - [1 - \exp\{-\lambda^{(b)}(2t-\tau-\theta)\}])$$
$$\times d\sigma(\tau)\,d\sigma(\theta).$$

Therefore

$$2AV = \frac{1}{E}\sigma^2 + 2\sigma\int_{-\infty}^{t}\psi(t-\tau)\,d\sigma(\tau)$$
$$- \int_{-\infty}^{t}\int_{-\infty}^{t}\psi(2t-\tau-\theta)\,d\sigma(\tau)\,d\sigma(\theta). \qquad (41)$$

From eq. (39)

$$\psi'(t) = \frac{d\psi(t)}{dt} = \sum_{b=1}^{N} \lambda^{(b)} B^{(b)} \exp(-\lambda^{(b)}t) H(t),$$

therefore, from eq. (36),

$$A\phi = \frac{1}{\eta}\sigma^2 + \int_{-\infty}^{t}\int_{-\infty}^{t} \psi'(2t-\tau-\theta)\,d\sigma(\tau)\,d\sigma(\theta). \qquad (42)$$

We now invert eq. (34) to find an equation for the stress in terms of the strain. If the two-sided Laplace transform

$$\bar{\kappa}(p) = \int_{-\infty}^{\infty} \exp(-pt)\,\kappa(t)\,dt \qquad (43)$$

is applied to eqs. (39) and (40),

$$\bar{\psi} = \sum_{b=1}^{N} \frac{B^{(b)}\lambda^{(b)}}{p(p+\lambda^{(b)})}, \qquad \bar{\epsilon} = \left(\frac{1}{E} + \frac{1}{\eta p} + p\bar{\psi}\right)\bar{\sigma} \quad (44)$$

or

$$\bar{\epsilon} = \left(\frac{1}{E} + \frac{1}{\eta p} + \sum_{b=1}^{N} \frac{B^{(b)}\lambda^{(b)}}{p+\lambda^{(b)}}\right)\bar{\sigma}. \quad (45)$$

Using the lemma and its corollaries in Section 5,

$$\bar{\sigma} = \left(E' + \eta' p + \sum_{r=1}^{N'} \frac{C^{(r)}p}{p+\mu^{(r)}}\right)\bar{\epsilon}. \quad (46)$$

The $C^{(r)}$ and $\mu^{(r)}$ are all real and positive. If $\frac{1}{E} = 0$, $\eta' > 0$; if $\frac{1}{E} > 0$, $\eta' = 0$; if $\frac{1}{\eta} = 0$, $E' > 0$; if $\frac{1}{\eta} > 0$, $E' = 0$. If $E' > 0$ and $\eta' > 0$, $N' = N - 1$; if $E' > 0$ and $\eta' = 0$, or if $E' = 0$ and $\eta' > 0$ then $N' = N$; if $E' = \eta' = 0$, $N' = N + 1$. If $\frac{1}{\eta} = 0$, $\lambda^{(r)} < \mu^{(r)} < \lambda^{(r+1)}$; if $\frac{1}{\eta} \neq 0$, $\lambda^{(r-1)} < \mu^{(r)} < \lambda^{(r)}$.

Inverting the transforms in eq. (46),

$$\sigma(t) = E'\epsilon(t) + \eta' D\epsilon(t) + \sum_{r=1}^{N'} \int_{-\infty}^{t} C^{(r)} \exp\{-\mu^{(r)}(t-\tau)\} d\epsilon(\tau). \quad (47)$$

The relaxation function $\chi(t)$ is defined as the stress required to produce a strain $H(t)$, excluding terms either that are infinite initially or that do not tend to zero as $t \to \infty$.

Putting $\epsilon(t) = H(t)$ in eq. (47)

$$\sigma(t) = E'H(t) + \eta'\delta(t) + \sum_{r=1}^{N'} C^{(r)} \exp(-\mu^{(r)}t)H(t),$$

where $\delta(t)$ is the Dirac delta function. Hence we have the result:

"The deviatoric and dilatational relaxation functions of a linear isotropic viscoelastic material, whose microscopic structure contains only a finite number of elements, are of the form

$$\chi(t) = \sum_{r=1}^{N'} C^{(r)} \exp(-\mu^{(r)}t)H(t), \quad (48)$$

where the $C^{(r)}$ and $\mu^{(r)}$ are positive constants."

Substituting back from eq. (48) into eq. (47)

$$\sigma(t) = E'\epsilon(t) + \eta' D\epsilon(t) + \int_{-\infty}^{t} \chi(t-\tau)\, d\epsilon(\tau). \qquad (49)$$

The stored and dissipated energies can be expressed in terms of the stress and strain or in terms of the strain and the relaxation function. Consider the last term in eq. (41):

$$\begin{aligned}
I &= \int_{-\infty}^{t}\int_{-\infty}^{t} \psi(2t-\tau-\theta)\, d\sigma(\tau)\, d\sigma(\theta) \\
&= \int_{-\infty}^{t} d\sigma(\tau)\left[\int_{-\infty}^{2t-\tau}\psi(2t-\tau-\theta)\, d\sigma(\theta) - \int_{t}^{2t-\tau}\psi(2t-\tau-\theta)\, d\sigma(\theta)\right] \\
&= \int_{-\infty}^{t} d\sigma(\tau)\int_{-\infty}^{2t-\tau}\psi(2t-\tau-\theta)\, d\sigma(\theta) \\
&\quad - \int_{t}^{\infty} d\sigma(\theta)\int_{-\infty}^{2t-\theta}\psi(2t-\tau-\theta)\, d\sigma(\tau),
\end{aligned}$$

on changing the order of integration in the last term. Using eq. (40)

$$\begin{aligned}
I = &\int_{-\infty}^{t} d\sigma(\tau)\left[\epsilon(2t-\tau) - \frac{1}{E}\sigma(2t-\tau) - \frac{1}{\eta}\int_{-\infty}^{2t-\tau}\sigma(\theta)\, d\theta\right] \\
&-\int_{t}^{\infty} d\sigma(\theta)\left[\epsilon(2t-\theta) - \frac{1}{E}\sigma(2t-\theta) - \frac{1}{\eta}\int_{-\infty}^{2t-\theta}\sigma(\tau)\, d\tau\right]. \quad (50)
\end{aligned}$$

Now $\int_{t}^{\infty}\sigma(2t-\theta)\, d\sigma(\theta) = -\int_{-\infty}^{t}\sigma(\tau)\, d\sigma(2t-\tau)$ on putting $2t-\theta=\tau$.

Therefore $-\int_{-\infty}^{t}\sigma(2t-\tau)\, d\sigma(\tau) + \int_{t}^{\infty}\sigma(2t-\theta)\, d\sigma(\theta)$
$$= -\int_{-\infty}^{t} d(\sigma(\tau)\sigma(2t-\tau)) = -\sigma^{2}.$$

And $-\int_{-\infty}^{t} d\sigma(\tau)\int_{-\infty}^{2t-\tau}\sigma(\theta)\, d\theta + \int_{t}^{\infty} d\sigma(\theta)\int_{-\infty}^{2t-\theta}\sigma(\tau)\, d\tau$
$$\begin{aligned}
&= -\left[\sigma(\tau)\int_{-\infty}^{2t-\tau}\sigma(\theta)\, d\theta\right]\bigg|_{\tau=-\infty}^{t} - \int_{-\infty}^{t}\sigma(\tau)\sigma(2t-\tau)\, d\tau \\
&\quad + \left[\sigma(\theta)\int_{-\infty}^{2t-\theta}\sigma(\tau)\, d\tau\right]\bigg|_{\theta=t}^{\infty} + \int_{t}^{\infty}\sigma(\theta)\sigma(2t-\theta)\, d\theta \\
&= -2\sigma(t)\int_{-\infty}^{t}\sigma(\theta)\, d\theta.
\end{aligned}$$

Therefore $I = \int_{-\infty}^{t}\epsilon(2t-\tau)\, d\sigma(\tau) - \int_{t}^{\infty}\epsilon(2t-\theta)\, d\sigma(\theta)$
$$- \frac{1}{E}\sigma^{2} - \frac{2}{\eta}\sigma\int_{-\infty}^{t}\sigma(\theta)\, d\theta.$$

40   THE THEORY OF LINEAR VISCOELASTICITY

Also $2\sigma \int_{-\infty}^{t} \psi(t-\tau) \, d\sigma(\tau) = 2\sigma \left(\epsilon - \frac{1}{E} \sigma - \frac{1}{\eta} \int_{-\infty}^{t} \sigma(\tau) \, d\tau \right).$

Substituting the last two equations in eq. (41),

$$2AV = 2\sigma\epsilon - \int_{-\infty}^{t} \epsilon(2t-\tau) \, d\sigma(\tau) + \int_{t}^{\infty} \epsilon(2t-\theta) \, d\sigma(\theta);$$

$$2AV = \sigma\epsilon + \int_{-\infty}^{t} \{\sigma(2t-\tau) \, d\epsilon(\tau) - \epsilon(2t-\tau) \, d\sigma(\tau)\} \quad (51)$$

on integrating the last term by parts. Equation (51) expresses the stored energy in terms of the stress and strain but can only be used directly when the stress and strain are known analytically at all times. A similar equation can be derived for the dissipation:

$$A\phi = \sigma D\epsilon + \int_{-\infty}^{t} \{\epsilon'(2t-\tau) \, d\sigma(\tau) - \sigma(2t-\tau) \, d\epsilon'(\tau)\}, \quad (52)$$

where $\epsilon'(\theta) = \dfrac{d\epsilon(\theta)}{d\theta}.$

If the stress is eliminated from eqs. (51) and (52) by means of eq. (49), it can be shown that

$$2AV = E'\epsilon^2 + \int_{-\infty}^{t}\int_{-\infty}^{t} \chi(2t-\tau-\theta) \, d\epsilon(\tau) \, d\epsilon(\theta) \quad (53)$$

and

$$A\phi = \eta'(D\epsilon)^2 - \int_{-\infty}^{t}\int_{-\infty}^{t} \chi'(2t-\tau-\theta) \, d\epsilon(\tau) \, d\epsilon(\theta), \quad (54)$$

where $\chi'(\theta) = \dfrac{d\chi(\theta)}{d\theta}.$

7. Sinusoidal oscillations, complex modulus and compliance

In this section a volume of viscoelastic material small enough to neglect spacial variations of stress is considered to be subject to a sinusoidally oscillating stress of frequency ω,

i.e. $\sigma_{ij} = R[\sigma_{ij}{}^0 e^{i\omega t}], \quad \sigma_{ij}{}^0$ complex constants, (55)

where $R[z]$ denotes the real part of z. After a sufficient time has elapsed for the effect of the initial conditions to be negligible, the strain ϵ_{ij} will also be of the form

$\epsilon_{ij} = R[\epsilon_{ij}{}^0 e^{i\omega t}], \quad \epsilon_{ij}{}^0$ complex constants. (56)

The normal co-ordinates ζ_i will also be sinusoidally oscillatory. Consequently the assumption made in integrating eq. (14) that $\zeta_i \to 0$ as $t \to -\infty$ is no longer valid, and therefore in analysing sinusoidal oscillations we must go back to eq. (14).

In eq. (14) substitute for σ_{ij} from eq. (55) and for ζ_i from
$$\zeta_i = R[\zeta_i{}^0 e^{i\omega t}], \quad \zeta_i{}^0 \text{ complex constants.} \tag{57}$$
Whence, since eq. (14) is valid at all times,
$$\zeta_i{}^0 = (1 - \Lambda^{(i)} + i\omega\kappa\Lambda^{(i)})^{-1} \sum_{m=1}^{3} \gamma_{jk}^{(i)} \sigma_{jl}{}^0 a_l^{(m)} a_k^{(m)}. \tag{58}$$

Substituting from eqs. (56) and (57) into eq. (11),
$$\epsilon_{ij}{}^0 = \sum_{k=1}^{M} \gamma_{ij}^{(k)} \zeta_k{}^0 = \sum_{k=1}^{m} (1 - \Lambda^{(i)} + i\omega\kappa\Lambda^{(i)})^{-1} \sum_{m=1}^{3} \gamma_{ij}^{(k)} \gamma_{pq}^{(k)} \sigma_{pl}{}^0 a_l^{(m)} a_q^{(m)}.$$

Substituting from eqs. (21), (23) and (15),
$$\epsilon_{ij}{}^0 = \alpha\sigma_{ij}{}^0 - \beta\delta_{ij}\sigma_{kk}{}^0 + \sum_{b=1}^{N} \frac{\lambda^{(b)}}{\lambda^{(b)} + i\omega} (\alpha^{(b)}\sigma_{ij}{}^0 - \beta^{(b)}\delta_{ij}\sigma_{kk}{}^0).$$

To this strain must be added that due to the relative molecular motion, i.e. from eq. (1)
$$e_{ij}{}^0 = \frac{1}{i\omega\eta} s_{ij}{}^0.$$

Separating the total strain into deviatoric and dilatational parts
$$e_{ij}{}^0 = \left(\alpha + \frac{1}{i\omega\eta} + \sum_{b=1}^{N} \frac{\lambda^{(b)} \alpha^{(b)}}{\lambda^{(b)} + i\omega}\right) s_{ij}{}^0 \tag{59}$$
and
$$\epsilon_{kk}{}^0 = \left(\gamma + \sum_{b=1}^{N} \frac{\lambda^{(b)} \gamma^{(b)}}{\lambda^{(b)} + i\omega}\right) \sigma_{kk}{}^0. \tag{60}$$

Again we shall consider
$$\epsilon^0 = \left(\frac{1}{E} + \frac{1}{i\omega\eta} + \sum_{b=1}^{N} \frac{\lambda^{(b)} B^{(b)}}{\lambda^{(b)} + i\omega}\right) \sigma^0, \tag{61}$$
which can be interpreted as any deviatoric component or as the dilatation.

The complex compliance $J(i\omega)$ is defined as ϵ^0/σ^0. Therefore
$$J(i\omega) = \frac{1}{E} + \frac{1}{i\omega\eta} + \sum_{b=1}^{N} \frac{\lambda^{(b)} B^{(b)}}{\lambda^{(b)} + i\omega}. \tag{62}$$

The complex modulus $Y(i\omega)$ is defined as σ^0/ϵ^0. Therefore

$$Y(i\omega) = (J(i\omega))^{-1} = \left(\frac{1}{E} + \frac{1}{i\omega\eta} + \sum_{b=1}^{N} \frac{\lambda^{(b)}B^{(b)}}{\lambda^{(b)} + i\omega}\right)^{-1}.$$

Using the lemma and corollaries of Section 5, and the notation of eqs. (45) and (46)

$$Y(i\omega) = E' + i\omega\eta' + \sum_{r=1}^{N'} \frac{C^{(r)}i\omega}{\mu^{(r)} + i\omega}. \tag{63}$$

Hence we have the result:

"The deviatoric and dilatational complex compliances and moduli of linear isotropic viscoelastic materials have alternating zeros and poles on the positive imaginary axis and neither zeros nor poles elsewhere. It is always a pole of a compliance which is nearest the origin and for the deviatoric compliance of a fluid this pole is situated at the origin."

Comparing eq. (45) with eq. (62) and eq. (46) with eq. (63), we see that

$$J(i\omega) = \frac{1}{E} + \frac{1}{i\omega\eta} + \overline{\psi'}^F \tag{64}$$

and

$$Y(i\omega) = E' + i\omega\eta' + \overline{\chi'}^F \tag{65}$$

where $\overline{\psi'}^F$ is the Fourier transform† of $d\psi(t)/dt$ and $\overline{\chi'}^F$ is the Fourier transform of $d\chi(t)/dt$. Equations (64) and (65) were first derived by B. Gross [5].

Let us now find the stored and dissipated energies in a sinusoidal oscillation. Substituting for ζ_i from eq. (57), using a bar to denote a complex conjugate,

$$\zeta_i^2 = \tfrac{1}{2}\zeta_i^0\overline{\zeta}_i^0 + \tfrac{1}{2}R[(\zeta_i^0)^2 \exp(2i\omega t)]. \tag{66}$$

Reference to eq. (12) shows that V will consist of two parts, the one a constant and the other oscillating sinusoidally with frequency 2ω. Now, using eq. (58),

$$\sum_{\Lambda^{(i)}=\Lambda^{(b)}} \zeta_i^0\overline{\zeta}_i^0(1-\Lambda^{(i)}) = (1-\Lambda^{(b)})\{(1-\Lambda^{(b)})^2$$
$$+ \omega^2\kappa^2(\Lambda^{(b)})^2\}^{-1} \sum_{\Lambda^{(i)}=\Lambda^{(b)}} \sum_{n=1}^{3}\sum_{m=1}^{3}\sigma_{jl}{}^0\sigma_{pr}{}^0\gamma_{jk}^{(i)}\gamma_{pq}^{(i)}a_l^{(m)}a_k^{(m)}a_q^{(n)}a_r^{(n)}.$$

† $\bar{x}^F(w) = \int_{-\infty}^{\infty} x(t)\exp(-iwt)\,dt,\ x(t) = \dfrac{1}{2\pi}\int_{-\infty}^{\infty}\bar{x}^F(w)\exp(iwt)\,dw.$

Using eqs. (15), (21) and (23)

$$\sum_{\Lambda^{(i)}=\Lambda^{(b)}} \zeta_i^0 \bar{\zeta}_i^0 (1 - \Lambda^{(i)}) = \frac{\lambda^{(i)2}}{\lambda^{(i)2} + \omega^2}$$
$$\times (\alpha^{(b)}\sigma_{ij}^0 \bar{\sigma}_{ij}^0 - \beta^{(b)}\sigma_{ii}^0 \bar{\sigma}_{jj}^0), \quad \Lambda^{(b)} > 0, \quad (67)$$

$$\sum_{\Lambda^{(i)}=\Lambda^{(b)}} \zeta_i^0 \bar{\zeta}_i^0 = \alpha\sigma_{ij}^0 \bar{\sigma}_{ij}^0 - \beta\sigma_{ii}^0 \bar{\sigma}_{jj}^0, \quad \Lambda^{(b)} = 0, \quad (68)$$

$$\sum_{\Lambda^{(i)}=\Lambda^{(b)}} R[(\zeta_i^0)^2 \exp(2i\omega t)] = R\left[\frac{\lambda^{(b)2}}{(\lambda^{(b)} + i\omega)^2} (\alpha^{(b)}\sigma_{ij}^0 \sigma_{ij}^0 - \beta^{(b)}\sigma_{ii}^0 \sigma_{jj}^0) \exp(2i\omega t)\right], \quad \Lambda^{(b)} > 0, \quad (69)$$

and
$$\sum_{\Lambda^{(i)}=\Lambda^{(b)}} R[(\zeta_i^0)^2 \exp(2i\omega t)] = R[\alpha\sigma_{ij}^0 \sigma_{ij}^0 - \beta\sigma_{ii}^0 \sigma_{jj}^0) \exp(2i\omega t)], \quad \Lambda^{(b)} = 0. \quad (70)$$

The stored energy can again be separated into deviatoric and dilatational parts and as before we shall write down a typical part.

The complex compliance $J(i\omega)$ can be separated into real and negative imaginary parts:

$$J(i\omega) = J_1(\omega) - iJ_2(\omega) \quad (71)$$

where $J_1(\omega)$ and $J_2(\omega)$ are real functions of ω. From eq. (62)

$$J_1(\omega) = \frac{1}{E} + \sum_{b=1}^{N} \frac{(\lambda^{(b)})^2 B^{(b)}}{(\lambda^{(b)})^2 + \omega^2} \quad (72)$$

and

$$J_2(\omega) = \frac{1}{\omega\eta} + \omega \sum_{b=1}^{N} \frac{\lambda^{(b)} B^{(b)}}{(\lambda^{(b)})^2 + \omega^2}. \quad (73)$$

From eqs. (12), (62) and (66) to (70)

$$4AV = J_1|\sigma^0|^2 + R[(J(i\omega) + i\omega J'(i\omega))(\sigma^0)^2 \exp(2i\omega t)], \quad (74)$$

where as before $A = 1$ for a deviatoric component, $A = 3$ for the dilatation and $J'(i\omega) = \frac{dJ(i\omega)}{d(i\omega)}$. If $\sigma^0 = |\sigma^0| \exp(i\alpha)$, from eqs. (71) to (74)

$$4AV = J_1|\sigma^0|^2 + |\sigma^0|^2(J_1 + \omega J_1'(\omega))\cos 2(\omega t + \alpha)$$
$$+ |\sigma^0|^2(J_2 + \omega J_2'(\omega))\sin 2(\omega t + \alpha). \quad (75)$$

Similarly it can be shown that

$$2A\phi = \omega J_2 |\sigma^0|^2 - \omega^2 R[J'(i\omega)(\sigma^0)^2 \exp(2i\omega t)] \quad (76)$$

or

$$2A\phi = \omega J_2 |\sigma^0|^2 - \omega^2 |\sigma^0|^2 J_2'(\omega) \cos 2(\omega t + \alpha) \\ + w^2 |\sigma^0|^2 J_1'(\omega) \sin 2(\omega t + \alpha). \quad (77)$$

Note that only the constant parts of the stored and dissipated energies in a sinusoidal oscillation are determined by the complex compliance at that frequency. To determine the variable parts, the derivative of the compliance must be known as well.

8. Operational form of the stress–strain equation

The stress–strain equation in the forms of eq. (24) was obtained by integrating eqs. (14) and then substituting for the ζ_i into eq. (11). The operational form is obtained by eliminating the ζ_i from eqs. (14) and (11) without first integrating eq. (14). The differential equations are linear with constant coefficients and therefore the operator D can be treated as an algebraic symbol in the subsequent manipulation until the equations are finally integrated. Comparing eqs. (58) with eqs. (14), the only difference is that $i\omega$ replaces D, and ζ_i^0 replaces ζ_i: Hence eq. (61) is an operational form of the stress–strain equation provided $i\omega$ is replaced by D and ϵ^0 and σ^0 by ϵ and σ respectively;

i.e. $$\epsilon = \left(\frac{1}{E} + \frac{1}{\eta D} + \sum_{b=1}^{N} \frac{\lambda^{(b)} B^{(b)}}{\lambda^{(b)} + D}\right)\sigma. \quad (78)$$

On cross-multiplication

$$E \prod_{b=1}^{N} (D + \lambda^{(b)}) D\epsilon = F_{N+1}(D)\sigma, \quad (79)$$

where F_{N+1} is a polynomial with constant coefficients of order $N + 1$ in D and with the coefficient of D^{N+1} equal to unity. But from eq. (78) by the lemma of Section 5,

$$\sigma = \left(\sum_{r=1}^{N+1} \frac{C^{(r)} D}{D + \mu^{(r)}}\right)\epsilon, \quad \text{if} \quad \frac{1}{E} \neq 0 \quad \text{and} \quad \frac{1}{\eta} \neq 0, \quad (80)$$

THREE-DIMENSIONAL LINEAR VISCOELASTICITY 45

and on cross-multiplication

$$\prod_{r=1}^{N+1}(D+\mu^{(r)})\sigma = F_N(D)\epsilon. \tag{81}$$

$F_N(D)$ is a polynomial with constant coefficients of order N in D and with the coefficient of D^N equal $\sum_{r=1}^{N+1} C^{(r)}$. Since (79) and (81) are the same equation, they are both identical to

$$E\prod_{b=1}^{N}(D+\lambda^{(b)})D\epsilon = \prod_{r=1}^{N+1}(D+\mu^{(r)})\sigma,$$
$$\frac{1}{E} \neq 0 \quad \text{and} \quad \frac{1}{\eta} \neq 0; \tag{82}$$

The equations for the other cases are

$$E\prod_{b=1}^{N}(D+\lambda^{(b)})\epsilon = \prod_{r=1}^{N}(D+\mu^{(r)})\sigma,$$
$$\frac{1}{E} \neq 0 \quad \text{and} \quad \frac{1}{\eta} = 0; \tag{83}$$

$$\eta'\prod_{b=1}^{N}(D+\lambda^{(b)})D\epsilon = \prod_{r=1}^{N}(D+\mu^{(r)})\sigma,$$
$$\frac{1}{E} = 0 \quad \text{and} \quad \frac{1}{\eta} \neq 0; \tag{84}$$

and $\quad \eta'\prod_{b=1}^{N+1}(D+\lambda^{(b)})\epsilon = \prod_{r=1}^{N}(D+\mu^{(r)})\sigma,$
$$\frac{1}{E} = 0 \quad \text{and} \quad \frac{1}{\eta} = 0. \tag{85}$$

All four cases can be summarized:

The stress–strain equation of a linear viscoelastic material can be expressed in the form

$$P\sigma = Q\epsilon, \tag{86}$$

where P and Q are polynomials with constant coefficients in the operator $D\left(\equiv \dfrac{d}{dt}\right)$. The zeros of the polynomials are all real, non-positive and they alternate. The least zero in absolute magnitude is a zero of Q. The ratio of any coefficient in P to any coefficient in Q is positive. The four cases then correspond to whether the least zero is $\left(\dfrac{1}{\eta} \neq 0\right)$ or is not $\left(\dfrac{1}{\eta} = 0\right)$ equal to zero and to whether the greatest zero in absolute magnitude is a zero of $P\left(\dfrac{1}{E} \neq 0\right)$ or a zero of $Q\left(\dfrac{1}{E} = 0\right)$.

Note that if σ and ϵ are varying sinusoidally, given by eqs. (55) and (56), then

$$J(i\omega) = \frac{P(i\omega)}{Q(i\omega)} \quad \text{and} \quad Y(i\omega) = \frac{Q(i\omega)}{P(i\omega)}, \tag{87}$$

where $P(i\omega)$ and $Q(i\omega)$ are the same functions of $i\omega$ as $P(D)$ and $Q(D)$ were of D.

The stored and dissipated energies can be expressed as homogeneous quadratic functions of the stress, of the strain and of their derivatives at the given time. Equation (35) can be written:

$$2AV = \frac{1}{E}\sigma^2 + \sum_{b=1}^{N} B^{(b)}(F^{(b)})^2 \tag{88}$$

where
$$F^{(b)} = \int_{-\infty}^{t} (1 - e^{-\lambda^{(b)}(t-\tau)})\, d\sigma(\tau). \tag{89}$$

Equation (34) in this notation is

$$\epsilon = \frac{1}{E}\sigma + \frac{1}{\eta}\int_{-\infty}^{t} \sigma(\tau)\, d\tau + \sum_{b=1}^{N} B^{(b)} F^{(b)}. \tag{90}$$

From eq. (89),

$$DF^{(b)} = \lambda^{(b)} \int_{-\infty}^{t} e^{-\lambda^{(b)}(t-\tau)}\, d\sigma(\tau). \tag{91}$$

Therefore

$$DF^{(b)} = \lambda^{(b)}\sigma - \lambda^{(b)} F^{(b)}. \tag{92}$$

If eq. (90) is differentiated with respect to time and if the $DF^{(b)}$ are eliminated using eq. (91),

$$D\epsilon = \frac{1}{E}D\sigma + \frac{1}{\eta}\sigma + \sum_{b=1}^{N} B^{(b)}\lambda^{(b)}\sigma - \sum_{b=1}^{N} B^{(b)}\lambda^{(b)} F^{(b)}.$$

Differentiate and substitute a further $n-1$ times:

$$D^n\epsilon = \frac{1}{E}D^n\sigma + \frac{1}{\eta}D^{n-1}\sigma + \sum_{m=0}^{n-1}\sum_{b=1}^{N} (-)^{n-m-1} B^{(b)}(\lambda^{(b)})^{n-m} D^m\sigma$$
$$+ (-)^n \sum_{b=1}^{N} B^{(b)}(\lambda^{(b)})^n\, F^{(b)}. \tag{93}$$

If $1/\eta = 0$, eq. (90) and eqs. (93) with $n = 1, 2, 3 \ldots N - 1$ are N equations which can be solved for the $NF^{(b)}$. The determinant of the coefficients of the $F^{(b)}$ is $\pm \prod\limits_{\substack{r,s=1 \\ r>s}}^{N} (\lambda^{(r)} - \lambda^{(s)})$, which is non-zero because the $\lambda^{(b)}$ are distinct. If $1/\eta \neq 0$, eqs. (93) with $n = 1, 2, 3 \ldots N$ are the N equations for the $NF^{(b)}$. Since the $B^{(b)}$ and the $\lambda^{(b)}$ can be expressed in terms of the coefficients of the polynomials P and Q, the $F^{(b)}$ are expressible as linear homogeneous functions of σ, of ϵ and of their time derivatives with coefficients which are functions of the coefficients of the polynomials P and Q. Substitution for the $F^{(b)}$ into eq. (88) gives V as a homogeneous quadratic function of σ, of ϵ and of their time derivatives, with coefficients that are functions of the coefficients of the polynomials P and Q.

From eqs. (36) and (92)

$$A\phi = \frac{1}{\eta}\sigma^2 + \sum_{b=1}^{N} \frac{B^{(b)}}{\lambda^{(b)}} (DF^{(b)})^2. \tag{94}$$

Differentiating eq. (90)

$$D\epsilon = \frac{1}{E}D\sigma + \frac{1}{\eta}\sigma + \sum_{b=1}^{N} B^{(b)} DF^{(b)}.$$

Substitute from eq. (92) for the $DF^{(b)}$ and differentiate. Repeat a further $n - 1$ times. This produces equations for the $DF^{(b)}$ analogous to eqs. (93) for the $F^{(b)}$. The $DF^{(b)}$, obtained by solving these equations, can be substituted into eq. (94). This expresses ϕ as a homogeneous quadratic function of σ, of ϵ and of their time derivatives with coefficients that are functions of the coefficients of the polynomials P and Q.

It can be seen, by comparison with eqs. (82) to (85), that the derivatives of σ and ϵ that occur in V and ϕ are all those derivatives that occur in $P\sigma = Q\epsilon$ except the highest-order derivative of both σ and ϵ. The only exceptions to this rule are for very simple forms of $P\sigma = Q\epsilon$, where $N = 0$ and the method of successive differentiation and substitution is not required. There are three exceptions:

$$\sigma = E\epsilon, \quad \sigma = \eta D\epsilon \quad \text{and} \quad \frac{1}{E}D\sigma + \frac{1}{\eta}\sigma = D\epsilon.$$

9. Model representation

It is often convenient to pictorially represent stress–strain laws by models. The two basic elements are the spring and the dashpot with respective stress–strain equations

$$\sigma = E\epsilon \quad \text{and} \quad \sigma = \eta D\epsilon.$$

The stress–strain law of any model, which must be a one-dimensional network of these elements is formed from the rule: "If two models are placed in series, the stress is the same in each and equal to the overall stress, but the overall strain is the sum of the strains of the separate models; if they are placed in parallel, the strain is the same in each and equal to the overall strain but the overall stress is the sum of the stresses in the separate models."

Note that these elements are not elements of the microscopic network previously considered. The latter lead to a force–extension equation, the former to a stress–strain equation. The model formed by placing the two basic elements in series is called a Maxwell element and has a stress–strain equation

$$D\epsilon = \frac{1}{E} D\sigma + \frac{1}{\eta} \sigma,$$

that formed by placing them in parallel is called a Voigt element and has equation

$$\sigma = E\epsilon + \eta D\epsilon.$$

The usefulness of the model representation is now apparent because from eqs. (78), (88) and (94) we see that the mechanical properties and the stored and dissipated energies of any linear viscoelastic material can be represented either by a set of Voigt elements in series or by a set of Maxwell elements in parallel.† The pictorial representation so obtained is frequently useful in describing the reaction of the material to imposed force or displacement. The stored and dissipated energies of the eight simplest models are given in Table 1.

Different authors use different nomenclature to describe the simple models. For example, the Voigt element is sometimes called the "Voigt–Kelvin" or even the "Kelvin"

† That the Maxwell elements in parallel can represent the stored and dissipated energies of the material is not shown above.

TABLE 1. The stored and dissipated energies for the eight simplest models

Stress-strain law	Model representation (in general, not unique)	Stored energy, V	Rate of dissipation, ϕ
$\sigma = E\epsilon$, elastic		$\frac{1}{2}E\epsilon^2 = \frac{1}{2E}\sigma^2$	0
$\sigma = \eta D\epsilon$, viscous		0	$\eta(D\epsilon)^2 = \frac{1}{\eta}\sigma^2$
$\frac{1}{E}D\sigma + \frac{1}{\eta}\sigma = D\epsilon$, Maxwell		$\frac{1}{2E}\sigma^2$	$\frac{1}{\eta}\sigma^2$
$\sigma = \eta D\epsilon + E\epsilon$, Voigt		$\frac{1}{2}E\epsilon^2$	$\eta(D\epsilon)^2$
$p_1 D\sigma + p_0\sigma = q_1 D\epsilon + q_0\epsilon$, three-element elastic		$\dfrac{p_1^2\sigma^2 - 2p_1 q_0\sigma\epsilon + q_0 q_1 (D\epsilon)^2}{2(p_0 q_1 - p_1 q_0)}$	$\dfrac{p_0^2\sigma^2 - 2p_0 q_0 \sigma\epsilon + q_0^2 \epsilon^2}{p_0 q_1 - p_1 q_0}$
$p_1 D\sigma + p_0\sigma = q_2 D^2\epsilon + q_1 D\epsilon$, three-element viscous		$\dfrac{p_1^2\sigma^2 - 2p_1 q_2 \sigma D\epsilon + q_2^2 (D\epsilon)^2}{2(p_0 q_1 - p_0 q_2)}$	$\dfrac{p_0 p_1 \sigma^2 - 2p_0 q_2 \sigma D\epsilon + q_1 q_2 (D\epsilon)^2}{p_1 q_1 - p_0 q_2}$
$p_2 D^2\sigma + p_1 D\sigma + p_0\sigma = q_2 D^2\epsilon + q_1 D\epsilon$, four element of first type		$\{(p_1^2 q_2 - p_0 p_2 q_2 + p_1 p_2 q_1)\sigma^2 + p_2^2 q_2 (D\sigma)^2 + q_1 q_2^2 (D\epsilon)^2 + 2p_2^2 p_1 q_2 \sigma D\sigma - 2p_2 q_2^2 D\sigma D\epsilon - 2q_2(p_1 q_2 - p_2 q_1)\sigma D\epsilon\} \times \{2(p_1 q_2 q_1 - p_2 q_1^2 - p_0 q_2^2)\}^{-1}$	$\{p_0(p_1 q_2 - p_2 q_1)\sigma^2 + q_1 q_2^2 (D\epsilon)^2 + p_2^2 q_1 (D\sigma)^2 + 2p_0 p_2 q_2 \sigma D\sigma - 2p_2 q_1 q_2 D\sigma D\epsilon - 2p_0 q_2^2 \sigma D\epsilon\} \times \{p_1 q_2 q_1 - p_2 q_1^2 - p_0 q_2^2\}^{-1}$
$p_1 D\sigma + p_0\sigma = q_2 D^2\epsilon + q_1 D\epsilon + q_0\epsilon$, four element of second type		$\{p_0 p_1^2 \sigma^2 + p_0 q_2^2 (D\epsilon)^2 + q_0 (p_1 q_1 - p_0 q_2)\epsilon^2 - 2p_0 p_1 q_2 D\sigma D\epsilon - 2p_1^2 q_0 \sigma\epsilon + 2p_1 q_0 q_2 \epsilon D\epsilon\} \times \{2(p_0 p_1 q_1 - p_0^2 q_2 - p_1^2 q_0)\}^{-1}$	$\{p_0^2 p_1 \sigma^2 + q_2(p_0 q_1 - p_1 q_0)(D\epsilon)^2 + p_1 q_0^2 \epsilon^2 - 2p_0^2 q_2 \sigma D\epsilon - 2p_0 p_1 q_0 \sigma\epsilon + 2p_0 q_0 q_2 \epsilon D\epsilon\} \times \{p_0 p_1 q_1 - p_0^2 q_2 - p_1^2 q_0\}^{-1}$

element. Jeffreys [6] calls a material which is elastic dilatational and Maxwell deviatoric "elasticoviscous", a material which is elastic dilatational and Voigt deviatoric "firmoviscous" and he uses a material, which is elastic dilatational and three-element elastic deviatoric, as a model for elastic after-working.

The models of Chapter 1 can be considered as particular cases of models representing one-dimensional stress–strain laws. If a rod, for which equation (37a) is valid, has cross-sectional area A and length L, then the tensile force on the rod is $F = \sigma_{11}A$ and the extension of the rod is $a = \epsilon_{11}L$. σ_{11} and ϵ_{11} can be eliminated from equation (37a) to give the force–extension equation of the rod, an equation of the type derived in Chapter 1. Because the rules for combining models in series and in parallel are the same in both cases, each model representation of this force–extension equation is the same as one representation of the stress–strain equation, except that the constants of the springs and dashpots in the former case are A/L times those in the latter. Conversely to each model of Chapter 1 there exist mechanically equivalent viscoelastic rods, whose stress–strain laws in tension and ratios of cross-sectional area to length are uniquely determined.

10. Retardation and relaxation spectra

It has been assumed in earlier sections that the number of joins inside each box, defined by Hypothesis 1, was finite and equal to n. Now let us investigate what happens as $n \to \infty$.

For any n however large the analysis of previous sections remains valid. There are four possible types of distribution of the $\lambda^{(b)}$ in the range $0 < \lambda^{(b)} < \infty$ as n tends to infinity: (i) the number of $\lambda^{(b)}$ remains finite—this case is identical to that for finite n; (ii) the number of $\lambda^{(b)}$ tends to infinity but there exists an $\epsilon > 0$ such that $|\lambda^{(b)} - \lambda^{(c)}| > \epsilon$ for any two $\lambda^{(b)}$ and $\lambda^{(c)}$; (iii) the number of $\lambda^{(b)}$ tends to infinity and, given any $\epsilon > 0$, for any $\lambda^{(b)}$ n can be chosen sufficiently large for there to exist a $\lambda^{(c)}$ such that $|\lambda^{(b)} - \lambda^{(c)}| < \epsilon$; (iv) the number of $\lambda^{(b)}$ tends to infinity such that the second condition of (ii) is valid over parts of the range $0 < \lambda^{(b)} < \infty$ and the second condition of (iii) is valid over the remainder of the range.

In case (iii) the infinite sum in eq. (39) is replaced by an integral when $n \to \infty$. It is conventional to call $\frac{1}{\lambda}$ the retardation time and to integrate with respect to it, i.e.

$$\psi(t) = \int_{\frac{1}{\lambda}=0}^{\infty} j\left(\frac{1}{\lambda}\right)\{1 - \exp(-\lambda t)\} d\left(\frac{1}{\lambda}\right) H(t); \qquad (95)$$

$j\left(\frac{1}{\lambda}\right)$ is known as the distribution function of retardation times or as the retardation spectrum. Similarly from eq. (48) in case (iii) when $n \to \infty$

$$\chi(t) = \int_{\frac{1}{\mu}=0}^{\infty} y\left(\frac{1}{\mu}\right) \exp(-\mu t) d\left(\frac{1}{\mu}\right) H(t); \qquad (96)$$

$\frac{1}{\mu}$ is known as the relaxation time and $y\left(\frac{1}{\mu}\right)$ as the distribution function of relaxation times or as the relaxation spectrum. $j\left(\frac{1}{\lambda}\right)$ and $y\left(\frac{1}{\mu}\right)$ must be real and non-negative for all real and non-negative values of their arguments.

In case (i) $j\left(\frac{1}{\lambda}\right)$ and $y\left(\frac{1}{\mu}\right)$ are finite sums of Dirac delta functions, in case (ii) they are infinite sums of delta functions, and in case (iv) they can be divided into two parts of which one part is a sum of delta functions. In all cases the eqs. (41), (53) and (75) for the stored energy and eqs. (42), (54) and (77) for the dissipation are still applicable. This section concludes with two examples, the one in case (ii) and the other in case (iii); the various viscoelastic functions will be found in each case.

Example 1

Consider an equation of the form

$$\frac{1}{\Gamma(D + \frac{1}{2})} \sigma = \frac{K}{\Gamma(D)} \epsilon, \qquad (97)$$

where Γ is the gamma function and K is a positive constant.

Since $\quad \dfrac{1}{\Gamma(z)} = z \exp(\gamma z) \prod_{n=1}^{\infty} \left\{\left(1 + \dfrac{z}{n}\right) \exp(-z/n)\right\},$

it is analytic for all finite values of z and has zeros at $z = 0, -1, -2, -3 \ldots$. Similarly $\frac{1}{\Gamma(z + \frac{1}{2})}$ is analytic for all finite values of z and has zeros at $z = -\frac{1}{2}, -\frac{3}{2}, -\frac{5}{2}, \ldots$. Both $\frac{1}{\Gamma(z)}$ and $\frac{1}{\Gamma(z + \frac{1}{2})}$ can be expanded as infinite series in integral powers of z. Putting $P = \frac{1}{\Gamma(D + \frac{1}{2})}$ and $Q = \frac{K}{\Gamma(D)}$, it is clear that eq. (97) represents the stress–strain law of a linear viscoelastic material because the conditions stated after eq. (86) are all satisfied. Since the least zero of Q in absolute magnitude is zero, one would expect the material to show long-term viscous flow; since neither P nor Q contain the greatest zero in absolute magnitude, the question of whether the material exhibits instantaneous elasticity is undecided. Equation (97) is an example of case (ii).

The creep function is defined as the strain response of the material to a stress $H(t)$, neglecting the instantaneous elasticity and the long-term viscous flow. Since both σ and ϵ are zero for $t < 0$, and since $\frac{1}{\Gamma(D + \frac{1}{2})}$ and $\frac{1}{\Gamma(D)}$ can be expanded as infinite series in integral powers of D, application of the two-sided Laplace transform to eq. (97) gives

$$\frac{1}{\Gamma(p + \frac{1}{2})} \bar{\sigma} = \frac{K}{\Gamma(p)} \bar{\epsilon}. \tag{98}$$

In this case $\bar{\sigma} = \frac{1}{p}$. Using the relation

$$\frac{\Gamma(z)\Gamma(a + 1)}{\Gamma(z + a)} = \sum_{n=0}^{\infty} \frac{(-)^n a(a - 1)(a - 2) \ldots (a - n)}{n!} \times \frac{1}{z + n}, \quad a > 0, \tag{99}$$

$$\bar{\epsilon} = \frac{\Gamma(p)}{Kp\Gamma(p + \frac{1}{2})} = \frac{1}{K\sqrt{\pi}} \sum_{n=0}^{\infty} \frac{(2n)!}{2^{2n}(n!)^2} \frac{1}{p(p + n)}.$$

Therefore $\quad \epsilon(t) = \frac{1}{K\sqrt{\pi}} \sum_{n=1}^{\infty} \frac{(2n)!}{2^{2n}(n!)^2 \cdot n}$

$$\times \{1 - \exp(-nt)\} H(t) + \frac{1}{K\sqrt{\pi}} t H(t)$$

and $\psi(t) = \dfrac{1}{K\sqrt{\pi}} \sum_{n=1}^{\infty} \dfrac{(2n)!}{2^{2n} \cdot (n!)^2 \cdot n} \{1 - \exp(-nt)\} H(t).$
(100)

The material exhibits long-term viscous flow of amount $\epsilon(t) = \dfrac{1}{\eta} H(t)$, where

$$\frac{1}{\eta} = \frac{1}{K\sqrt{\pi}} \qquad (101)$$

but no instantaneous elastic response since $\epsilon(t) = 0$ at $t = 0$, hence $1/E = 0$.

To find the relaxation function put $\bar{\epsilon} = \dfrac{1}{p}$ in eq. (98). If eq. (99) is used,

$$\bar{\sigma} = \frac{K\Gamma(p + \tfrac{1}{2})}{p\Gamma(p)} = \frac{K\Gamma(p + \tfrac{1}{2})}{\Gamma(p + 1)}$$

$$= \frac{K}{\sqrt{\pi}} \sum_{n=0}^{\infty} \frac{(2n)!}{2^{2n}(n!)^2} \frac{1}{p + \tfrac{1}{2} + n},$$

and $\sigma(t) = \dfrac{K}{\sqrt{\pi}} \sum_{n=0}^{\infty} \dfrac{(2n)!}{2^{2n} \cdot (n!)^2} \exp\{-(n + \tfrac{1}{2})t\} H(t).$

Hence $\chi(t) = \dfrac{K}{\sqrt{\pi}} \sum_{n=0}^{\infty} \dfrac{(2n)!}{2^{2n} \cdot (n!)^2} \exp\{-(n + \tfrac{1}{2})t\} H(t)$ (102)

and $\eta' = E' = 0.$ (103)

For a material possessing only a finite number of $\lambda^{(b)}$, $\dfrac{1}{E} = 0$ implied $\eta' \neq 0$. This is not the case for this material. But it can be shown that $\chi(t) \to \infty$ as $t \to +0$. Hence there is a singularity in the stress response of this material to a strain $H(t)$ but the singularity is no longer of the delta function type.

The complex compliance and complex modulus are respectively

$$J(i\omega) = \frac{\Gamma(i\omega)}{K\Gamma(i\omega + \tfrac{1}{2})} \quad \text{and} \quad Y(i\omega) = \frac{K\Gamma(i\omega + \tfrac{1}{2})}{\Gamma(i\omega)}.$$
(104)

Example 2

Consider a creep function of the form

$$\psi(t) = \frac{t^\nu}{K\Gamma(1+\nu)} H(t) \text{ where } 0 \leqslant \nu \leqslant 1 \text{ and } K > 0. \tag{105}$$

If $\nu = 0$, this is an elastic solid; if $\nu = 1$, it is a viscous fluid. To show eq. (105) is the creep function of a linear viscoelastic material, it is sufficient to show that there exists a corresponding retardation spectrum which is non-negative. If eq. (95) is differentiated with respect to t,

$$\frac{d\psi(t)}{dt} = \int_0^\infty \frac{1}{\lambda} j\left(\frac{1}{\lambda}\right) \exp(-\lambda t) \, d\lambda \, H(t).$$

It can be seen that $\frac{1}{\lambda} j\left(\frac{1}{\lambda}\right)$ is the inverse one-sided Laplace transform (not p-multiplied) of $\frac{d\psi(t)}{dt}$. Differentiating eq. (105) with respect to t,

$$\frac{d\psi(t)}{dt} = \frac{t^{\nu-1}}{K\Gamma(\nu)} H(t).$$

Hence
$$\frac{1}{\lambda} j\left(\frac{1}{\lambda}\right) = \frac{\lambda^{-\nu}}{\Gamma(1-\nu)} \frac{1}{K\Gamma(\nu)},$$

$$j\left(\frac{1}{\lambda}\right) = \frac{\lambda^{1-\nu}}{K\Gamma(1-\nu)\Gamma(\nu)}. \tag{106}$$

$j\left(\frac{1}{\lambda}\right) > 0$ for $0 < \lambda < \infty$ and therefore eq. (105) is the creep function of a linear viscoelastic material. It is an example of case (iii).

If we suppose the material to exhibit neither instantaneous elasticity nor long term viscous flow, then the stress–strain equation, by eq. (40), can be written in the form

$$\epsilon(t) = \frac{1}{K\Gamma(1+\nu)} \int_{-\infty}^t (t-\tau)^\nu \, d\sigma(\tau),$$

or, on integration by parts,

$$\epsilon(t) = \frac{1}{K\Gamma(\nu)} \int_{-\infty}^t \sigma(\tau)(t-\tau)^{\nu-1} \, d\tau.$$

But the fractional integral $D^{-n}f(t)$ is defined by [7]

$$D^{-n}f(t) = \int_{-\infty}^{t} \frac{(t-\tau)^{n-1}}{\Gamma(n)} f(\tau) \, d\tau,$$

hence
$$\epsilon(t) = \frac{1}{K} D^{-\nu}\sigma(t)$$

or
$$\sigma = KD^{\nu}\epsilon. \tag{107}$$

This type of stress–strain law has been suggested by G. W. Scott-Blair.

An alternative form of eq. (107) is

$$\sigma(t) = KD^{-(1-\nu)}D\epsilon = \frac{K}{\Gamma(1-\nu)} \int_{-\infty}^{t} (t-\tau)^{-\nu} d\epsilon(\tau). \tag{108}$$

Hence the relaxation function is

$$\chi(t) = \frac{K}{\Gamma(1-\nu)} t^{-\nu} H(t) \tag{109}$$

and, by an argument similar to that used to derive eq. (106), the relaxation spectrum is

$$y\left(\frac{1}{\mu}\right) = \frac{K}{\Gamma(\nu)\Gamma(1-\nu)} \mu^{\nu+1}. \tag{110}$$

The complex modulus and compliance are

$$Y(i\omega) = K(i\omega)^{\nu} \tag{111}$$

and
$$J(i\omega) = \frac{1}{K} (i\omega)^{-\nu}. \tag{112}$$

Note that $\arg Y(i\omega) = -\arg J(i\omega) = \dfrac{\nu\pi}{2}$ is independent of ω.

11. Summary of the results of Chapter 2

The properties of a linear viscoelastic material are developed from the hypothesis that the microscopic structure of such a material is mechanically equivalent to a network of elastic and viscous elements.

The stored energy and the rate of dissipation of energy can be found for any material element at any time. For an isotropic material, each deviatoric component of strain is related solely to the corresponding deviatoric component of stress and the dilatational part of the strain solely to the dilatational part of the stress; the energies are the sum of the respective energies for the deviatoric components and for the dilatation. It is shown both that the strain can be expressed in terms of the stress increments and the creep function, and that the stress can be expressed in terms of the strain increments and the relaxation function. The complex compliances and moduli have alternating zeros and poles on the positive imaginary axes and no zeros or poles elsewhere. The stress–strain law can be expressed in operational form, $P\sigma = Q\epsilon$ where P and Q are polynomials in d/dt with constant coefficients. The zeros of P and Q are all real and nonpositive and they alternate.

The energies can be expressed in terms of either the creep-function and the stress at previous times, or the relaxation function and the strain at previous times, or the stress, strain and their time derivatives at a given time or, for a sinusoidal oscillation, in terms of the complex compliance and its derivative with respect to frequency. It is shown that models consisting of Voigt element in series or of Maxwell elements in parallel can represent the mechanical properties and the stored and dissipated energies of any viscoelastic material.

The analysis can be extended to networks containing an infinite number of elements. Two examples, one of each two different cases, are given.

REFERENCES

1. D. R. BLAND: *Proc. Roy. Soc.* 1959, A **250**, 524.
2. W. L. BRAGG and M. LOMAR: *Proc. Roy. Soc.* 1949, A **196**, 171.
3. H. JEFFREYS: *Cartesian tensors* (p. 70, equation 20, Cambridge University Press, 1931).
4. D. R. BLAND: *Rheologica Acta*, to be published.
5. B. GROSS: *J. Appl. Phys.*, 1948, **19**, 257.
6. H. JEFFREYS: The Earth, Third Edition (Cambridge University Press, 1952, section 1.04).
7. D. V. WIDDER: The Laplace Transform (p. 71, Princeton University Press, 1946).

CHAPTER 3

STRESS ANALYSIS I: SINUSOIDAL OSCILLATION PROBLEMS

1. Stress analysis in viscoelasticity

The last chapter was concerned with the stress–strain equation satisfied by a viscoelastic material. This chapter and the following two are concerned with finding the stress and displacement fields inside viscoelastic bodies when acted on by external forces and/or restraints. This is the subject of stress analysis.

Stress analysis in continua mechanics is concerned with the simultaneous solution of three sets of equations subject to given boundary conditions. The first set comes from the analysis of strain, which is based solely on geometrical considerations, and are known as either (i) the strain–displacement equations, which for small strains are

$$\epsilon_{ij} = \tfrac{1}{2}(u_{j,i} + u_{i,j}) \tag{1}$$

where u_i is the displacement or (ii) the strain rate–velocity equations, which are

$$D\epsilon_{ij} = \tfrac{1}{2}(v_{j,i} + v_{i,j}) \tag{2}$$

where $v_i = Du_i$ is the velocity. D now denotes a partial time derivative of a tensor with its components referred to axes fixed in a material element.

The second set comes from the analysis of stress, which depends on geometry and on Newton's laws, and are known as the equations of motion (for a continuous medium). They are

$$\sigma_{ij,j} + \rho X_i = \rho a_i \tag{3}$$

where ρ is the density, X_i the body force and a_i the acceleration. For linear viscoelastic solids the displacement u_i is small enough to enable a_i to be replaced by $\partial^2 u_i/\partial t^2$, and D by $\partial/\partial t$; the equations of motion become

$$\sigma_{ij,j} + \rho X_i = \rho \frac{\partial^2 u_i}{\partial t^2}. \tag{4}$$

These approximations are made in Chapters 3–5. The results so obtained will also be valid for fluids provided the displacement is small.

The third set are the stress–strain equations. These are different for different materials and it is the particular form of these equations which characterizes the different branches of continua mechanics. For a particular problem in viscoelasticity convenience dictates which of the different but equivalent forms of the stress–strain equations, developed in Chapter 2, is chosen.

This chapter treats problems in which the dependent variables vary sinusoidally with time. The following chapter treats problems in which the dependent variables vary sufficiently slowly with time to enable the inertia terms in the equations of motion, i.e. the right-hand sides of the equations (3) or (4) to be neglected. In the fifth chapter the dependent variables vary rapidly with time.

Experimentally it is possible to measure with reasonable accuracy the complex moduli for frequencies greater than about 10^{-6} sec^{-1} and the creep and relaxation functions for times greater than about 1 sec—a review of experimental techniques has been given by Ferry [1]. Consequently it is convenient to use the complex modulus or compliance form of the stress–strain equations when the dependent variables vary either sinusoidally, or rapidly with time, and the creep or relaxation function form when the variation is very slow. When the final result of the analysis is evaluated for actual materials, the physical properties of the material will be expressed in a form for which the numerical values are known in the time range under consideration.

2. Propagation of sinusoidally oscillating waves in an infinite medium

We look for solutions of eqs. (1) and (4) for a viscoelastic material in which all the dependent variables vary sinusoidally with time. We assume the displacement is sufficiently small to enable the acceleration term to be expressed as $\rho \partial^2 u_i/\partial t^2$. Put

$$\sigma_{ij} = R[\bar{\sigma}_{ij} \exp(i\omega t)], \epsilon_{ij} = R[\bar{\epsilon}_{ij} \exp(i\omega t)], \text{etc.}, \qquad (5)$$

where $\bar{\sigma}_{ij}$, $\bar{\epsilon}_{ij}$, etc. are functions, in general complex, of the spacial co-ordinates only. It is assumed that any body force is also sinusoidally oscillatory with radian frequency ω. By definition of the complex moduli, see Chapter 2, Section 7,

$$\bar{s}_{ij} = Y_s \bar{e}_{ij} \qquad (6)$$

and

$$\bar{\sigma}_{kk} = Y_\nu \bar{\epsilon}_{kk}, \qquad (7)$$

where Y_s and Y_ν are the deviatoric and dilatational complex moduli respectively. Equations (1) and (4) become

$$\bar{\epsilon}_{ij} = \tfrac{1}{2}(\bar{u}_{j,i} + \bar{u}_{i,j}) \qquad (8)$$

and

$$\bar{\sigma}_{ij,j} + \rho \bar{X}_i + \rho \omega^2 \bar{u}_i = 0. \qquad (9)$$

From eqs. (6) and (7)

$$\bar{\sigma}_{ij} = \bar{s}_{ij} + \tfrac{1}{3}\bar{\sigma}_{kk}\delta_{ij} = Y_s \bar{e}_{ij} + \tfrac{1}{3}Y_\nu \bar{\epsilon}_{kk}\delta_{ij}$$
$$= Y_s \bar{\epsilon}_{ij} + \tfrac{1}{3}(Y_\nu - Y_s)\bar{\epsilon}_{kk}\delta_{ij}.$$

Substitute for $\bar{\epsilon}_{ij}$ from eq. (8):

$$\bar{\sigma}_{ij} = \tfrac{1}{2}Y_s(\bar{u}_{j,i} + \bar{u}_{i,j}) + \tfrac{1}{3}(Y_\nu - Y_s)\bar{u}_{k,k}\delta_{ij}. \qquad (10)$$

Substitute for $\bar{\sigma}_{ij}$ into eq. (9):

$$\tfrac{1}{2}Y_s \bar{u}_{i,jj} + \tfrac{1}{6}(2Y_\nu + Y_s)\bar{u}_{j,ji} + \rho \bar{X}_i + \rho \omega^2 \bar{u}_i = 0. \qquad (11)$$

If eq. (11) is differentiated with respect to x_i,

$$\tfrac{1}{3}(2Y_s + Y_\nu)\bar{u}_{j,jii} + \rho \omega^2 \bar{u}_{j,j} = 0 \qquad (12)$$

whenever

$$X_{i,i} = 0. \qquad (13)$$

Equation (12) is satisfied by

$$\bar{u}_{j,j} = A \exp\left\{\pm i\omega \sqrt{\left(\frac{3\rho}{2Y_s + Y_r}\right)} A_j x_j\right\}, \qquad (14)$$

where A is a constant and A_i is a constant vector of unit modulus. Therefore†

$$u_{j,j} = R\left[A \exp\left(i\omega\left\{t \pm \sqrt{\left(\frac{3\rho}{2Y_s + Y_\nu}\right)}A_j x_j\right\}\right)\right] \quad (15)$$

is a solution of eqs. (1) and (4). $u_{j,j}$ is the dilatation. Equations (15) and (13) state that plane waves of dilatation are propagated through a viscoelastic medium, whenever the divergence of the sinusoidally oscillating body force is zero, with velocity v_D where

$$v_D = \left\{R\left[\sqrt{\left(\frac{3\rho}{2Y_s + Y_\nu}\right)}\right]\right\}^{-1}, \quad (16)$$

and with attenuation α_D where

$$\alpha_D = -\omega I\left[\sqrt{\left(\frac{3\rho}{2Y_s + Y_\nu}\right)}\right]. \quad (17)$$

Differentiate eq. (11) with respect to x_k, interchange i and k and subtract:

$$\tfrac{1}{2}Y_s(\bar{u}_{i,kjj} - \bar{u}_{k,ijj}) + \rho\omega^2(\bar{u}_{i,k} - \bar{u}_{k,i}) = 0 \quad (18)$$

whenever

$$X_{i,k} - X_{k,i} = 0. \quad (19)$$

Equation (18) is satisfied by

$$\tfrac{1}{2}(\bar{u}_{k,i} - \bar{u}_{i,k}) = B_{ik} \exp\left\{\pm i\omega\sqrt{\left(\frac{2\rho}{Y_s}\right)}B_j x_j\right\}, \quad (20)$$

where B_{ik} is a constant anti-symmetric tensor and B_j a constant vector of unit modulus. Therefore

$$\tfrac{1}{2}(u_{k,i} - u_{i,k}) = R\left[B_{ik}\exp\left(i\omega\left\{t \pm \sqrt{\left(\frac{2\rho}{Y_s}\right)}B_j x_j\right\}\right)\right] \quad (21)$$

is a solution of eqs. (1) and (4). $\tfrac{1}{2}(u_{k,i} - u_{i,k})$ is the rotation. Equations (21) and (19) state that plane waves of rotation are propagated through a viscoelastic medium, whenever the

† $R[z]$ and $I[z]$ denote the real and imaginary parts of z respectively.

curl of the sinusoidally oscillating body force is zero, with velocity v_R where

$$v_R = \left\{ R\left[\sqrt{\frac{2\rho}{Y_s}}\right] \right\}^{-1}, \qquad (22)$$

and with attenuation α_R, where

$$\alpha_R = -\omega I\left[\sqrt{\frac{2\rho}{Y_s}}\right]. \qquad (23)$$

From eq. (22) it is seen that the rotational wave velocity is independent of frequency only if $R[Y_s^{-\frac{1}{2}}]$ is constant. From eq. (2.63) it is seen that this is only true when Y_s is a real constant. In this case the material is elastic for deviatoric stress, $Y_s = 2\mu$ where μ is the shear modulus, and the velocity is $\sqrt{(\mu/\rho)}$. Furthermore from eqs. (23) and (2.63) it follows that the attenuation is only zero when the material is elastic for deviatoric stress and that for all other materials the attenuation varies with frequency.

"All viscoelastic materials, except for those materials which have an elastic deviatoric stress–strain relation, are dispersive and dissipative to sinusoidal rotational waves, and the attenuation varies with frequency."

Since $2Y_s + Y_\nu$ cannot be real unless both Y_s and Y_ν are real, it follows from equations (16) and (17) that

"All viscoelastic materials, except for purely elastic materials, are dispersive and dissipative to sinusoidal dilatational waves and the attenuation varies with frequency."

The actual displacements associated with the plane waves of dilatation or of rotation have yet to be found. A plane sinusoidal wave has displacements of the form

$$u_j = R[C_j \exp\{i\omega(t \pm Zn_k x_k)\}] \qquad (24)$$

or

$$\bar{u}_j = C_j \exp(\pm i\omega Z n_k x_k), \qquad (25)$$

where C_j is a constant vector, possibly complex, Z is a complex constant and n_k is the unit vector in the direction of propagation of the wave. Substitute from equation (25) into equation (11):

$$-\tfrac{1}{2}C_i\omega^2 Z^2 Y_s - \tfrac{1}{6}\omega^2 Z^2(2Y_\nu + Y_s)C_j n_j n_i + \rho\omega^2 C_i = 0,$$

whenever
$$X_i = 0,$$

i.e.
$$C_j n_j n_i = 3 \frac{2\rho Z^{-2} - Y_s}{2Y_v + Y_s} C_i.$$

This is a vector equation and therefore either the two vectors are equal in magnitude and direction or the two vectors both have zero coefficients. In the former case

$$C_i \parallel n_i, \quad Z^2 = \frac{3\rho}{2Y_s + Y_v}$$

and
$$u_j = R\left[A n_j \exp\left(i\omega\left\{t \pm \sqrt{\left(\frac{3\rho}{2Y_s + Y_v}\right)} n_k x_k\right\}\right)\right], \quad (26)$$

where A is a complex constant; and in the latter case

$$C_i \perp n_i, \quad Z^2 = \frac{2\rho}{Y_s}$$

and
$$u_j = R\left[C_j \exp\left(i\omega\left\{t \pm \sqrt{\left(\frac{2\rho}{Y_s}\right)} n_k x_k\right\}\right)\right]. \quad (27)$$

The displacement u_j in eq. (26) satisfies $u_{j,i} - u_{i,j} = 0$ and $u_{i,i} \neq 0$. The displacement u_j in eq. (27) satisfies $u_{j,j} = 0$ and $u_{j,i} - u_{i,j} \neq 0$. The results of this paragraph can be summarized as

"The only plane sinusoidal waves possible in a linear viscoelastic material under no sinusoidal body force are a wave of dilatation and no rotation and a wave of rotation and no dilatation. In the former the motion is along the direction of propagation, and in the latter the motion is along any direction normal to the direction of propagation."

This result is identical to that for an elastic medium.

The stress associated with either wave can be found by substitution from eq. (25) into eq. (10):

$$\bar{\sigma}_{ij} = \{\tfrac{1}{2}Y_s(C_j n_i + C_i n_j) + \tfrac{1}{3}(Y_v - Y_s)C_l n_l \delta_{ij}\} i\omega Z \exp(\pm i\omega Z n_k x_k).$$

Hence for the dilatational plane wave

$$\sigma_{ij} = -I\Bigg[A(Y_s n_i n_j + \tfrac{1}{3}(Y_\nu - Y_s)\delta_{ij})\omega\sqrt{\frac{3\rho}{2Y_s + Y_\nu}}$$

$$\times \exp\left(i\omega\left\{t \pm \sqrt{\left(\frac{3\rho}{2Y_s + Y_\nu}\right)n_k x_k}\right\}\right)\Bigg]; \quad (28)$$

and for the rotational plane wave

$$\sigma_{ij} = -I\Bigg[\tfrac{1}{2}Y_s(C_j n_i + C_i n_j)\omega\sqrt{\left\{\frac{2\rho}{Y_s}\right\}}$$

$$\times \exp\left(i\omega\left\{t \pm \sqrt{\left(\frac{2\rho}{Y_s}\right)n_k x_k}\right\}\right)\Bigg] \quad (29)$$

where $C_i n_i = 0$.

The rotational wave velocity becomes infinite as $\omega \to \infty$ if $Y_s \to \infty$ or $J_s \to 0$ as $\omega \to \infty$. From eqs. (2.62) it is seen that this is the case for a material whose basic network contains only a finite number of elements and which does not exhibit instantaneous elasticity when subject to deviatoric stress. The dilatational wave velocity becomes infinite as $\omega \to \infty$ if either or both of Y_s and $Y_\nu \to \infty$ as $\omega \to \infty$.

For dispersive but non-dissipative media the concept of group velocity arises from consideration of the superposition of two waves of equal amplitude and nearly equal frequency. For a viscoelastic material, excepting the elastic case, two such waves can have an equal amplitude at all times only at one particular point because they are attenuated by different amounts. Consequently the concept of group velocity in its original form cannot be applied. However the importance of group velocity in non-dissipative media is that it is equal to the velocity of propagation of energy, sometimes called the "energy velocity". The energy velocity is defined as the time average rate of energy flow across unit area, normal to the direction of wave propagation, divided by the time average of the energy density. This concept can be taken over to dissipative media provided that the energy density is understood as the sum of the kinetic and the stored energies per unit volume and specifically excludes all energy dissipated in that volume prior to the current time.

We now calculate the energy velocity for both plane waves. The rate of work across unit area in the direction of propagation is $-\sigma_{ij}n_j Du_i$. The time average rate of energy flow is therefore equal to $\frac{1}{2}R[\bar{\sigma}_{ij}n_j(D\bar{u}_i)^*]$ where a star denotes a complex conjugate.† For the dilatational wave from eqs. (28) and (26),

$$\sigma_{ij}n_j = R\left[i\omega A\sqrt{\left(\frac{\rho(2Y_s+Y_\nu)}{3}\right)}n_i\right.$$
$$\left.\times \exp\left(i\omega\left\{t-\sqrt{\left(\frac{3\rho}{2Y_s+Y_\nu}\right)}n_k x_k\right\}\right)\right]$$

and

$$Du_i = R\left[i\omega A n_i \exp\left(i\omega\left\{t-\sqrt{\left(\frac{3\rho}{2Y_s+Y_\nu}\right)}n_k x_k\right\}\right)\right]. \quad (30)$$

Therefore

$$\frac{1}{2}R[\bar{\sigma}_{ij}n_j(D\bar{u}_i)^*] = \frac{\sqrt{\rho}}{2\sqrt{3}}\omega^2|A|^2$$
$$\times \exp\left\{-2\omega I\left[\sqrt{\left(\frac{3\rho}{2Y_s+Y_\nu}\right)}\right]n_k x_k\right\}R[\sqrt{2Y_s+Y_\nu}] \quad (31)$$

for the dilatational wave.

For the rotational wave from eqs. (29) and (27),

$$\sigma_{ij}n_j = R\left[\tfrac{1}{2}i\omega Y_s C_i\sqrt{\frac{2\rho}{Y_s}}\exp\left(i\omega\left\{t-\sqrt{\left(\frac{2\rho}{Y_s}\right)}n_k x_k\right\}\right)\right]$$

and

$$Du_i = R\left[i\omega C_i \exp\left(i\omega\left\{t-\sqrt{\left(\frac{2\rho}{Y_s}\right)}n_k x_k\right\}\right)\right]. \quad (32)$$

Therefore

$$\frac{1}{2}R[\bar{\sigma}_{ij}n_j(D\bar{u}_i)^*] = \frac{\sqrt{2\rho}}{4}\omega^2|C_i|^2\exp\left\{-2\omega I\left[\sqrt{\left(\frac{2\rho}{Y_s}\right)}\right]n_k x_k\right\}$$
$$\times R[\sqrt{Y_s}]. \quad (33)$$

The time average of the energy density, $\bar{E}$, is equal to the sum of the time average of the kinetic energy per unit volume plus the time average of the stored energy per unit volume. Using eqs. (2.31) and (2.74),

$$\bar{E} = \tfrac{1}{4}\rho|D\bar{u}_i|^2 + \tfrac{1}{4}R[J_s]|s_{ij}|^2 + \tfrac{1}{12}R[J_\nu]|\sigma_{kk}|^2.$$

† For a proof of this result see, e.g. [2].

SINUSOIDAL OSCILLATION PROBLEMS

For the dilatational wave from eqs. (30) and (28)

$$\bar{E} = \tfrac{1}{4}\rho\omega^2|A|^2 \exp\left\{-2\omega I\left[\sqrt{\left(\frac{3\rho}{2Y_s + Y_\nu}\right)}\right]n_k x_k\right\}$$
$$\times \left(1 + R[J_s]|Y_s|^2 \frac{2}{|2Y_s + Y_\nu|} + R[J_\nu]|Y_\nu|^2 \frac{1}{|2Y_s + Y_\nu|}\right).$$

Now $J_s = Y_s^{-1}$,

therefore $R[J_s] = |Y_s|^{-2} R[Y_s]$.

Hence the energy velocity for a dilatational wave

$$= \frac{2}{\sqrt{3\rho}} R[\sqrt{(2Y_s + Y_\nu)}]\left(1 + \frac{R[2Y_s + Y_\nu]}{|2Y_s + Y_\nu|}\right)^{-1}$$
$$= \left\{R\left[\sqrt{\left(\frac{3\rho}{2Y_s + Y_\nu}\right)}\right]\right\}^{-1} \quad \text{by the lemma below.†}$$

For the rotational wave from eqs. (32) and (29),

$$\bar{E} = \tfrac{1}{4}\rho\omega^2|C_i|^2 \exp\left\{-2\omega I\left[\sqrt{\frac{2\rho}{Y_s}}\right]n_k x_k\right\}(1 + R[J_s]|Y_s|).$$

Hence the energy velocity for a rotational wave

$$= \sqrt{\frac{2}{\rho}} R[\sqrt{Y_s}]\left(1 + \frac{R[Y_s]}{|Y_s|}\right)^{-1} = \left\{R\left[\sqrt{\frac{2\rho}{Y_s}}\right]\right\}^{-1}.$$

Comparison of the equations for the energy velocities with eqs. (16) and (22) shows:

"The energy velocities for dilatational and rotational waves in a linear viscoelastic material are equal to the respective wave velocities."

† LEMMA: $R[\sqrt{z}]\left(1 + \dfrac{R[z]}{|z|}\right)^{-1} = \left\{R\left[\dfrac{2}{\sqrt{z}}\right]\right\}^{-1}.$

Proof: Put $z = re^{i\theta}$ where r and θ are real.

Then L.H.S. $= \sqrt{r}\cos\dfrac{\theta}{2}(1 + \cos\theta)^{-1}$

$= \sqrt{r}\cos\dfrac{\theta}{2}\left(2\cos^2\dfrac{\theta}{2}\right)^{-1} = \tfrac{1}{2}\sqrt{r}\sec\dfrac{\theta}{2}.$

and R.H.S. $= \{R[2r^{-1/2}e^{-i\theta/2}]\}^{-1} = \tfrac{1}{2}\sqrt{r}\sec\dfrac{\theta}{2}.$

For the elastic material, which is non-dispersive and non-dissipative, this reduces to the well-known result that the wave, group and energy velocities are all equal.

In vector notation eq. (12) becomes

$$\tfrac{1}{3}(2Y_s + Y_\nu)\nabla^2(\text{div } \bar{\mathbf{u}}) + \rho\omega^2 \text{ div } \bar{\mathbf{u}} = 0. \qquad (34)$$

This equation has a spherically symmetric solution,

$$\text{div } \bar{\mathbf{u}} = \frac{A}{r}\exp\left\{\pm i\omega\sqrt{\left(\frac{3\rho}{2Y_s + Y_\nu}\right)}r\right\}, \quad \begin{array}{l}A \text{ constant} \\ \text{assumed real,}\end{array}$$

whence

$$\text{div } \mathbf{u} = \frac{A}{r} R\left[\exp\left(i\omega\left\{t + \sqrt{\left(\frac{3\rho}{2Y_s + Y_\nu}\right)}r\right\}\right)\right]. \qquad (35)$$

This solution represents spherical dilatational waves sinusoidal in time spreading out from (or converging towards) a point source. Their velocity is $\left\{R\left[\sqrt{\dfrac{3\rho}{2Y_s + Y_\nu}}\right]\right\}^{-1}$ and their amplitude at radius r is

$$\frac{A}{r}\exp\left\{-\omega I\left[\sqrt{\left(\frac{3\rho}{2Y_s + Y_\nu}\right)}\right]r\right\}.$$

Equation (34) has a solution in cylindrical co-ordinates (r, θ, z) dependent only on r, which is of the form

$$\text{div } \bar{\mathbf{u}} = AH_0^{(2)}\left(\sqrt{\frac{3\rho}{2Y_s + Y_\nu}}\,\omega r\right), \quad A \text{ constant,}$$

where $H_0^{(2)}(z)$ is a Hankel function of order zero, whence

$$\text{div } \mathbf{u} = R\left[AH_0^{(2)}\left(\sqrt{\frac{3\rho}{2Y_s + Y_\nu}}\,\omega r\right)\exp(i\omega t)\right]. \qquad (36)$$

For large z, $\quad H_0^{(2)}(z) \sim \sqrt{\dfrac{2}{\pi z}}\exp\left\{-i\left(z - \dfrac{\pi}{4}\right)\right\}.$

Therefore for large r, $\text{div } \mathbf{u} \sim R\left[A\exp\left(i\dfrac{\pi}{4}\right)\sqrt{\dfrac{2(2Y_s + Y_\nu)}{3\rho\pi\omega r}}\right.$

$$\left. \times \exp\left(i\omega\left\{t - \sqrt{\left(\frac{3\rho}{2Y_s + Y_\nu}\right)}r\right\}\right)\right].$$

This solution represents right circular cylindrical waves, sinusoidal in time, spreading out from a uniform line source. Convergent waves are given by the solution involving the function $H_0^{(1)}(z)$. It is left as an exercise for the reader to determine the displacement and the stress associated with these dilatational waves. The solution for a time dependent uniform pressure acting on the surface of a spherical cavity in an infinite medium has been given by Berry [3]. He also solved, in the same paper, the problem of forced torsional oscillations in a circular cylinder.

Equation (18) in vector notation is

$$\tfrac{1}{2} Y_s \nabla^2 (\mathrm{curl}\ \bar{\mathbf{u}}) + \rho \omega^2 (\mathrm{curl}\ \bar{\mathbf{u}}) = 0, \qquad (37)$$

where $\nabla^2 \equiv \mathrm{grad}\ \mathrm{div} - \mathrm{curl}\ \mathrm{curl}$. Solutions for the three components of $\mathrm{curl}\ \bar{\mathbf{u}}$ can be found in various co-ordinate systems; the divergence of the vector so found must always be zero. No spherically symmetric rotational wave is possible.

3. The correspondence principle for sinusoidal oscillations

In the last section solutions were found for the stress and displacement starting from eqs. (5) to (9). These equations differ from the equations for an elastic material only in that the complex moduli have replaced the elastic moduli. Hence we can obtain the solution to certain sinusoidal viscoelastic problems by a "correspondence principle": "If the elastic solution for any dependent variable in a particular problem is of the form $f = R[\bar{f}_E \exp(i\omega t)]$ and if the elastic moduli in $\bar{f}_E$ are replaced by the corresponding complex moduli to give $\bar{f}_{VE}$ then the viscoelastic solution for that variable in the corresponding problem is given by $f = R[\bar{f}_{VE} \exp(i\omega t)]$." By "corresponding problem" is meant the identical problem except that the body concerned is viscoelastic instead of elastic. The principle can only be used if (i) the elastic solution is known, (ii) no operation in obtaining the elastic solution would have a corresponding operation in the viscoelastic solution which would involve separating either complex modulus into real and imaginary parts, with the exception of the final determination of f from $\bar{f}$ and (iii) the boundary conditions for the two cases are identical. An example, where condition (ii) would render the principle invalid, is the determination of the maximum of $|f|$ with respect to ω. It would be legitimate to use the principle to find f for the viscoelastic case but, since the determination of the magnitude of a complex quantity involves distinction between its real and imaginary parts, the determination of the maximum of $|f|$ in the two cases must be treated separately.

Equations (14) and (21) in the last section could have been obtained from the elastic solutions by the correspondence principle. However the treatment of energy velocity involves the use of complex conjugates and therefore the principle could not have been applied there. Approximate solutions for elastic bodies from the strength of materials can be adapted for viscoelastic bodies under the same boundary conditions. Such solutions will only be valid under the conditions under which the elastic solution are valid. Examples of this technique will now be given.

4. The vibrating reed

The reed is a rectangular parallellepiped of viscoelastic material with thickness small compared to length and breadth. One end (of the length) of the reed is clamped and is forced to perform vibrations of constant amplitude and frequency in the direction of the thickness. The other end is free and the problem is either to determine its motion in terms of the complex modulus at the imposed frequency or to determine the modulus from the observed motion. Mathematically these two problems are identical, one needs to find the equation of motion of the free end.

If x is distance measured along the reed from the clamped end and if v denotes the transverse displacement, then for the elastic reed performing sinusoidal vibrations of radian frequency ω

$$\frac{d^4\bar{v}}{dx^4} - \frac{m\omega^2}{EI}\bar{v} = 0, \tag{38}$$

where m is the mass per unit length, E Young's modulus and I the moment of inertia of the cross-section. The boundary conditions are $\bar{v} = V$, $\dfrac{d\bar{v}}{dx} = 0$ at $x = 0$ and $\dfrac{d^2\bar{v}}{dx^2} = \dfrac{d^3\bar{v}}{dx^3} = 0$ at $x = l$ since the bending moment and shearing force are zero at a free end. The solution of eq. (38) for these boundary conditions gives the displacement at the free end as

$$\bar{v} = V \frac{\cosh \psi + \cos \psi}{1 + \cosh \psi \cos \psi},$$

where

$$\psi = \left(\frac{m\omega^2}{EI}\right)^{1/4} l. \tag{39}$$

If A denotes the ratio of the amplitudes of oscillation of the free to the clamped end and θ the phase lag of the free end behind the clamped end,

$$Ae^{-i\theta} = \frac{\cosh \psi + \cos \psi}{1 + \cosh \psi \cos \psi}. \tag{40}$$

The solution for a viscoelastic material is given by the correspondence principle as eq. (39) and (40) provided that the elastic modulus in eq. (39) is replaced by the corresponding viscoelastic complex modulus.

The viscoelastic complex moduli corresponding to the various elastic moduli will now be found in terms of Y_s and Y_v. Using eqs. (6) and (7)

$$\bar{\epsilon}_{ij} = \bar{e}_{ij} + \tfrac{1}{3}\bar{\epsilon}_{kk}\delta_{ij} = \frac{1}{Y_s}\bar{s}_{ij} + \frac{1}{3Y_v}\bar{\sigma}_{kk}\delta_{ij}$$

$$= \frac{1}{Y_s}\bar{\sigma}_{ij} - \frac{1}{3}\left(\frac{1}{Y_s} - \frac{1}{Y_v}\right)\bar{\sigma}_{kk}\delta_{ij}.$$

In particular

$$\bar{\epsilon}_{11} = \frac{1}{3}\left(\frac{2}{Y_s} + \frac{1}{Y_v}\right)\bar{\sigma}_{11} - \frac{1}{3}\left(\frac{1}{Y_s} - \frac{1}{Y_v}\right)(\bar{\sigma}_{22} + \bar{\sigma}_{33}).$$

For an elastic material

$$\bar{\epsilon}_{11} = \frac{1}{E}\bar{\sigma}_{11} - \frac{\nu}{E}(\bar{\sigma}_{22} + \bar{\sigma}_{33}).$$

Comparing the last two equations, it is seen that the viscoelastic complex modulus Y_T corresponding to the Young's modulus for an elastic solid is given by

$$Y_T = 3\left(\frac{2}{Y_s} + \frac{1}{Y_\nu}\right)^{-1}, \qquad (41)$$

and the viscoelastic Poisson's ratio ν by

$$\nu = \left(\frac{1}{Y_s} - \frac{1}{Y_\nu}\right)\left(\frac{2}{Y_s} + \frac{1}{Y_\nu}\right)^{-1} = \frac{Y_\nu - Y_s}{2Y_\nu + Y_s}. \qquad (42)$$

TABLE 3.1.
Corresponding Elastic and Viscoelastic Moduli

Modulus	Shear	Compression	Young's	Poisson's ratio	Lamé's constant
Elastic	μ	k	E	ν	$\lambda = k - \frac{2}{3}\mu$
Viscoelastic	$\frac{1}{2}Y_s$	$\frac{1}{3}Y_\nu$	$3\left(\frac{2}{Y_s}+\frac{1}{Y_\nu}\right)^{-1}=Y_T$	$\dfrac{Y_\nu - Y_s}{2Y_\nu + Y_s} = \nu$	$\frac{1}{3}(Y_\nu - Y_s)$

The amplitude ratio A and the phase lag θ between the free and clamped ends of a viscoelastic reed are therefore given by

$$A e^{-i\theta} = \frac{\cosh \psi + \cos \psi}{1 + \cosh \psi \cos \psi} \qquad (40)$$

where

$$\psi = \left\{\frac{m\omega^2}{3I}\left(\frac{2}{Y_s} + \frac{1}{Y_\nu}\right)\right\}^{1/4}. \qquad (43)$$

To determine the maximum(s) of the amplitude ratio A with respect to frequency, A must be found explicitly from eqs. (40) and (43). Details of the method are given by Bland and Lee [4].

5. Free radial vibrations of a solid sphere†

The governing equations for the radial vibrations of an elastic sphere in spherical polar co-ordinates are

$$\epsilon_{rr} = \frac{\partial u}{\partial r}, \quad \epsilon_{\theta\theta} = \epsilon_{\phi\phi} = \frac{u}{r}, \quad \frac{\partial \sigma_{rr}}{\partial r} + \frac{2}{r}(\sigma_{rr} - \sigma_{\theta\theta}) = \rho \frac{\partial^2 u}{\partial t^2},$$

$$\sigma_{rr} = \lambda\left(\frac{\partial u}{\partial r} + 2\frac{u}{r}\right) + 2\mu \frac{\partial u}{\partial r}, \quad \sigma_{\theta\theta} = \lambda\left(\frac{\partial u}{\partial r} + 2\frac{u}{r}\right) + 2\mu \frac{u}{r}.$$

where u is the displacement in the radial direction and λ and μ are Lamé's constants.

Solving for u,

$$(\lambda + 2\mu)\left(\frac{\partial^2 u}{\partial r^2} + \frac{2}{r}\frac{\partial u}{\partial r} - \frac{2}{r^2}u\right) = \rho \frac{\partial^2 u}{\partial t^2}.$$

The solution to this equation for radian frequency ω which is finite at $r = 0$ is

$$u = R\left[\left(\frac{\kappa}{r}\cos \kappa r - \frac{1}{r^2}\sin \kappa r\right)\exp(i\omega t)\right], \tag{44}$$

where

$$\kappa^2 = \frac{\rho \omega^2}{\lambda + 2\mu}. \tag{45}$$

No external force acts on the surface of the sphere, i.e. $\sigma_{rr} = 0$ when $r = a$. Hence the characteristic equation for the frequency ω is

$$a\kappa \cot a\kappa = 1 - \frac{\lambda + 2\mu}{4\mu}(a\kappa)^2. \tag{46}$$

All the operations used to obtain this solution were linear so the correspondence principle can be applied to solve the corresponding viscoelastic problem. Using Table (3.1), $\lambda + 2\mu$ must be replaced by $\frac{1}{3}(Y_v + 2Y_s)$ and μ by $\frac{1}{2}Y_s$. Equations (45) and (46) become

$$\kappa^2 = \frac{3\rho\omega^2}{Y_v + 2Y_s} \tag{47}$$

and

$$a\kappa \cot a\kappa = 1 - \frac{Y_v + 2Y_s}{6Y_s}(a\kappa)^2. \tag{48}$$

† The solution for the elastic sphere is due to Poisson.

If Y_s/Y_ν is a real constant, independent of frequency, then the roots $a\kappa$ of eq. (48) are real, otherwise they are complex. In either case on substitution for κ into eq. (47) it is seen that ω will be complex (except in the special case of an elastic solid). The eigenfrequencies will always be complex in a sinusoidal free oscillation problem for a viscoelastic material because the material dissipates energy and no work is done on the material by external forces.

Let us investigate further the solution in a particular case. An elastic material is said to satisfy "Poisson's condition" if $\nu = \tfrac{1}{4}$. The same definition is adopted for a viscoelastic material. From Table (3.1), it is seen that this implies $2Y_\nu = 5Y_s$ for all ω. The particular case chosen will be for a Maxwell material satisfying Poisson's condition, with

$$\tfrac{1}{5}Y_\nu = \tfrac{1}{2}Y_s = \frac{E}{1 + (E/i\omega\eta)} \tag{49}$$

and $\dfrac{E}{\omega\eta} \ll 1$ for the frequencies to be considered.

Equation (48) becomes

$$a\kappa \cot a\kappa = 1 - \tfrac{3}{4}(a\kappa)^2.$$

The first five roots of this equation are $a\kappa/\pi = 0\cdot 8160$, $1\cdot 9285$, $2\cdot 9359$, $3\cdot 9658$ and $4\cdot 9728$. Equation (47) gives ω as

$$\omega = \frac{\kappa}{\sqrt{3\rho}}\sqrt{Y_\nu + 2Y_s}$$

$$= \kappa\sqrt{\frac{3E}{\rho}}\left(1 + \frac{E}{i\omega\eta}\right)^{-1/2}.$$

Using $\dfrac{E}{\omega\eta} \ll 1$,

$$\omega = \kappa\sqrt{\frac{3E}{\rho}}\left(1 + \frac{i}{2}\frac{\sqrt{\rho E}}{\sqrt{3}\kappa\eta}\right) = \kappa\sqrt{\frac{3E}{\rho}} + \frac{iE}{2\eta}.$$

Substituting back into eq. (44), the radial displacement is

$$u = A\left(\frac{\kappa}{r}\cos\kappa r - \frac{\sin\kappa r}{r^2}\right)\exp\left(-\frac{E}{2\eta}t\right)\cos\kappa\sqrt{\left(\frac{3E}{\rho}\right)}t. \tag{50}$$

The time taken for the displacement to be reduced to $1/e$ of its initial value is $2\eta/E$ which is independent both of the frequency and of the radius of the sphere. In this case of a Maxwell material with $E/\omega\eta \ll 1$, it is seen that the displacement is everywhere equal to that for an elastic solid times a damping factor $\exp\left(-\dfrac{E}{2\eta}t\right)$ when both materials satisfy Poisson's condition.

6. Rayleigh waves

Rayleigh waves are propagated parallel to the plane surface of a semi-infinite viscoelastic medium. If the x-axis is taken in the direction of propagation and the z-axis along a normal to the surface so that $z = 0$ is the surface and $z > 0$ is in the medium, then the components of displacement in the x and z directions, u and w respectively, in an elastic medium are given by (see e.g. [5])

$$u = AR[-i\kappa\{\exp(-qz) - 2qs(s^2 + \kappa^2)^{-1}\exp(-sz)\} \\ \times \exp\{i(\omega t - \kappa x)\}] \quad (51)$$

and $\quad w = AR[q\{\exp(-qz) - 2\kappa^2(s^2 + \kappa^2)^{-1}\exp(-sz)\}$
$$\times \exp\{i(\omega t - \kappa x)\}], \quad (52)$$

where
$$\left. \begin{array}{c} q^2 = \kappa^2(1 - \alpha\theta^2), \\ s^2 = \kappa^2(1 - \theta^2), \quad \theta^2 = \dfrac{\rho\omega^2}{\mu\kappa^2}, \quad \alpha = \dfrac{\mu}{\lambda + 2\mu}, \end{array} \right\} \quad (53)$$

θ^2 satisfies

$$\theta^6 - 8\theta^4 + (24 - 16\alpha)\theta^2 + 16\alpha - 16 = 0, \quad (54)$$

and A is a real constant.

The displacements for a Rayleigh wave in a viscoelastic material are given by the correspondence principle and eqs. (51) to (54). Using Table (3.1) $\alpha = \mu/(\lambda + 2\mu)$ is replaced by $3Y_s/(2Y_v + 4Y_s)$ and μ by $\tfrac{1}{2}Y_s$, eq. (54) is solved for θ^2, eqs. (53) for κ^2, q^2 and s^2 and finally substitution of these quantities into eqs. (51) and (52) gives the actual displacements. The root of eq. (54) chosen must lead to positive real parts of both q and s. It has not yet been shown that for any viscoelastic material there is one and only one such root.

Note that in this section, as in Section 2, where the body is of infinite extent the damping term is a term in x. In the last section, where the body was finite in extent, the damping term was a term in t.

An interesting application of this theory is the damping of Rayleigh waves when they are transmitted, as a result of earthquakes, through the outer layers of the earth's surface.[†] Ewing and Press [7] give the velocity of Rayleigh waves of periods from 250 to 350 sec as 4·8 km/sec and the attenuation, when allowance has been made for the curvature of the earth's surface, as 0.8×10^{-5} km^{-1}. The attenuation is very small and so the earth acts nearly as an elastic material. To account for the small amount of damping, it will be assumed that the earth is a Maxwell material with $E/\omega\eta \ll 1$ in the frequency range considered. For simplicity it is further assumed that the earth satisfies Poisson's condition so that the complex moduli are given by eq. (49). We try to find values of E and η, which lead to the observed values of velocity and attenuation.

$$\alpha = 3Y_s/(2Y_p + 4Y_s) = \tfrac{1}{3} \quad \text{by eq. (49).}$$

Equation (54) becomes $(\theta^2 - 4)(3\theta^4 - 12\theta^2 + 8) = 0$, whence $\theta^2 = 4, 2 \pm \tfrac{2}{3}\sqrt{3}$. For an elastic solid the relevant root is $2 - \tfrac{2}{3}\sqrt{3}$. If it is assumed that there is no discontinuous change in the type of wave propagated as $\eta \to \infty$, the relevant root for the Maxwell material is $2 - \tfrac{2}{3}\sqrt{3}$, i.e. $\theta = \sqrt{2 - \tfrac{2}{3}\sqrt{3}} = 0{\cdot}9194$.

From eqs. (49) and (53) and Table (3.1),

$$\kappa = \frac{\omega}{\theta}\sqrt{\frac{2\rho}{Y_s}} = \frac{\omega}{\theta}\sqrt{\frac{\rho}{E}}\left(1 - \frac{iE}{2\omega\eta}\right);$$

whence, from either eq. (51) or (52), the velocity of the Rayleigh wave $= \dfrac{\omega}{R[\kappa]} = \theta\sqrt{\left(\dfrac{E}{\rho}\right)}$ and the attenuation of the Rayleigh wave $= -I[\kappa] = \dfrac{\sqrt{(\rho E)}}{2\theta\eta}$.

[†] A review of viscoelastic effects in the earth has been written by B. Gutenberg [6].

Substituting the experimental values of velocity and attenuation, $\theta = 0\cdot 9194$; and, assuming $\rho = 6$,

$$E = 1\cdot 56 \times 10^{12} \text{ dyn/cm}^2, \quad \eta = 2\cdot 08 \cdot 10^{16} \text{ dyn sec/cm}^2. \quad (55)$$

A mean value of ω is $1/50$ sec^{-1}, hence $E/\omega\eta = 0\cdot 37 \times 10^{-2}$. The neglect of squares of $E/\omega\eta$ is therefore justified.

Thus if the earth's crust were formed of a Maxwell material satisfying Poisson's condition and with constants given by eq. (55) the observed values of Rayleigh wave velocity and attenuation would obtain. Since $2\eta/E = 2\cdot 67 \times 10^4$ sec it is clear that the damping in the earth's crust is sufficient to prevent the radial vibrations considered in the last section from building up in the earth.

REFERENCES

1. J. D. FERRY: *Rheology*, Vol. 2, 433. Ed. F. R. EIRICH, Academic Press, New York, 1956.
2. L. BRILLOUIN: *Wave propagation in periodic structures*, Second Edition (Section 19, pp. 70–72, Dover Publications, 1953).
3. D. S. BERRY: *J. Mech. Phys. Solids* **6** (1958) 177.
4. D. R. BLAND and E. H. LEE: *J. Appl. Phys.* **26** (1955) 1497.
5. H. KOLSKY: *Stress waves in solids* (p. 16 *et seq.*, Clarendon Press, 1954).
6. B. GUTENBERG: *Rheology*, Vol. 2 (Chapter 2), Rheological problems of the earth's interior. Ed. F. R. EIRICH, Academic Press, New York, 1956.
7. M. EWING and F. PRESS: *Bull. Seismol. Soc. Amer.* **44** (1954) 477.

CHAPTER 4

STRESS ANALYSIS II: QUASI-STATIC PROBLEMS

1. The correspondence principle

A quasi-static problem is one in which the dependent variables vary sufficiently slowly with respect to time to enable the inertia terms in the equations of motion to be neglected. The governing equations for small strains are

$$\epsilon_{ij} = \tfrac{1}{2}(u_{j,i} + u_{i,j}), \tag{1}$$
$$\sigma_{ij,j} + \rho X_i = 0 \tag{2}$$

and the viscoelastic stress–strain relationships. It is possible to measure the complex-moduli for frequencies as low as 10^{-6} per sec and so it is not practical to use the complex moduli for problems in which the range of the time variable exceeds about 10^6 sec or about 10 days. Creep and relaxation functions can be measured at any time starting about 1 second after application of the stress or strain respectively. Since many applications involve applying forces to viscoelastic materials (e.g. creep experiments) for several weeks, a method of solution is employed which expresses the stress–strain relation in a form derivable from creep or relaxation measurements. The Laplace transform of these measurements is the form chosen because it always converges and, when a numerical evaluation is required, only one integral is needed for each value of the transform variable p.

Equations (2.45) and (2.46) give the Laplace transforms of the stress–strain equation as†

$$\bar{\epsilon} = \left(\frac{1}{E} + \frac{1}{\eta p} + \sum_{b=1}^{N} \frac{B^{(b)}\lambda^{(b)}}{p + \lambda^{(b)}}\right)\bar{\sigma}$$

† A bar over a variable in this chapter denotes a Laplace transform.

or
$$\bar{\sigma} = \left(E' + \eta'p + \sum_{r=1}^{N'} \frac{C^{(r)}p}{p + \mu^{(r)}}\right)\bar{\epsilon}.$$

If these equations are compared with eqs. (2.62) and (2.63),
$$\bar{\epsilon} = J(p)\bar{\sigma} \tag{3}$$
and
$$\bar{\sigma} = Y(p)\bar{\epsilon} \tag{4}$$

where $J(p)$ and $Y(p)$ are the same functions of p as $J(i\omega)$ and $Y(i\omega)$ were of $i\omega$. It follows from eqs. (2.44) and (2.48) that

$$J(p) = \frac{1}{E} + \frac{1}{\eta p} + p\bar{\psi}(p) \tag{5}$$
and
$$Y(p) = E' + \eta'p + p\bar{\chi}(p). \tag{6}$$

Hence $\frac{1}{p}J(p)$ and $\frac{1}{p}Y(p)$ are the Laplace transforms of the strain response to a stress $H(t)$ and of the stress response to a strain $H(t)$ of the given viscoelastic material, i.e. of the quantities measured in creep and relaxation tests respectively. We shall refer to $J(p)$ and $Y(p)$ as "p-varying compliance" and "p-varying modulus" respectively.

The deviatoric and dilatational forms of eqs. (3) and (4) are

$$\bar{e}_{ij} = J_s(p)\bar{s}_{ij} \quad \text{and} \quad \bar{\epsilon}_{ii} = J_v(p)\bar{\sigma}_{ii} \tag{7}$$
and
$$\bar{s}_{ij} = Y_s(p)\bar{e}_{ij} \quad \text{and} \quad \bar{\sigma}_{ii} = Y_v(p)\bar{\epsilon}_{ii}. \tag{8}$$

The Laplace transforms of eqs. (1) and (2) are

$$\bar{\epsilon}_{ij} = \tfrac{1}{2}(\bar{u}_{j,i} + \bar{u}_{i,j}) \tag{9}$$
and
$$\bar{\sigma}_{ij,j} + \rho\bar{X}_i = 0. \tag{10}$$

The Laplace transforms used so far have been from $-\infty$ to ∞. In many actual problems the stress and the strain are zero for $t < 0$ and only become non-zero for $t > 0$. In such cases, by analogy with the terminology used in the theory of electrical transmission lines, we speak of the viscoelastic material as being "initially dead". For such problems the Laplace transform from $-\infty$ to $+\infty$ is replaced with one from 0 to $+\infty$. Equations (3) to (10) are unaltered because σ_{ij} and ϵ_{ij} are zero for $t < 0$ and $\psi(t)$ and $\chi(t)$ contain a factor $H(t)$ which is zero for $t < 0$. For "initially dead" problems three boundary conditions will be required everywhere on the surface of the material for all times $t > 0$.

If the Laplace transforms of the governing equations and of the boundary conditions[†] exist for an initially dead quasi-static problem for a viscoelastic material, and if they are compared with those for the same problem for an elastic material, it can be seen that they differ only in that the elastic moduli are replaced by the corresponding p-varying moduli. Consequently the solution to a quasi-static viscoelastic problem can be found from the solution of the corresponding elastic problem, provided that the Laplace transforms of the boundary conditions exist, by a correspondence principle:

"In the elastic solution replace the dependent variables and the boundary conditions by their Laplace transforms and the elastic moduli by the corresponding p-varying moduli. Inversion of the expressions so obtained for the transforms of the dependent variables gives the viscoelastic solution for these variables."

Table (3.1) can still be used for corresponding moduli provided that Y_s and Y_v are now interpreted as $Y_s(p)$ and $Y_v(p)$. In the following three sections particular problems will be solved by this method.

2. Expansion of a reinforced cylinder by internal pressure[‡]

A thick right-circular cylinder of viscoelastic material is surrounded by a band of elastic material and both are maintained in plane strain. The cylinder is subject to internal pressure $\Pi(t)$ and the outside of the band is stress free. The

[†] It is assumed that the boundary conditions are of the form: stress or displacement given.

[‡] Due to W. B. Woodward and J. R. M. Radok [1].

internal and external radii of the cylinder are a and b respectively and the thickness of the band is h where $h \ll b$. The problem is to find the stress distribution within the cylinder. We shall first find the solution for an elastic cylinder and then use the correspondence principle for a viscoelastic cylinder.

Take plane polar co-ordinates r, θ in any plane normal to the generators with the origin on the axis of the cylinder. Let F_θ be the circumferential tension per unit length in the band. Then for equilibrium of the band

$$F_\theta = -b(\sigma_{rr})_{r=b} \tag{11}$$

since $-(\sigma_{rr})_{r=b}$ is the interfacial pressure. Since $b \gg h$ the circumferential strain in the band is nearly equal to $F_\theta(1 - \nu_B^2)/hE_B$ where ν_B and E_B are the elastic constants of the band. The circumferential strain in the cylinder at the interface is $\{\sigma_{\theta\theta} - \nu\sigma_{rr}/(1-\nu)\}_{r=b}(1-\nu^2)/E$ where ν and E are the elastic constants of the cylinder. For continuity the two strains must be equal. Hence, if F_θ is eliminated by eq. (11),

$$\sigma_{rr} = \beta \sigma_{\theta\theta} \quad \text{at} \quad r = b \tag{12}$$

where

$$\beta = \frac{1 - \nu^2}{\nu(1 + \nu) - (1 - \nu_B^2)bE/hE_B}. \tag{13}$$

The solution for the stresses σ_{rr} and $\sigma_{\theta\theta}$ inside an elastic cylinder is

$$\sigma_{rr} = A - B/r^2 \quad \text{and} \quad \sigma_{\theta\theta} = A + B/r^2,$$

where A and B are constants. They are determined by $\sigma_{rr} = -\Pi$ at $r = a$ and by eq. (12).

Hence

$$\sigma_{rr} = -\Pi \frac{\beta\{(b^2/r^2) + 1\} - \{(b^2/r^2) - 1\}}{\beta\{(b^2/a^2) + 1\} - \{(b^2/a^2) - 1\}} \tag{14}$$

and

$$\sigma_{\theta\theta} = \Pi \frac{\beta\{(b^2/r^2) - 1\} - \{(b^2/r^2) + 1\}}{\beta\{(b^2/a^2) + 1\} - \{(b^2/r^2) - 1\}}. \tag{15}$$

THE THEORY OF LINEAR VISCOELASTICITY

Equations (14) and (15) give the stress components σ_{rr} and $\sigma_{\theta\theta}$ in an elastic cylinder. To obtain the viscoelastic solution, first apply the Laplace transform to these equations:

$$\bar{\sigma}_{rr} = -\bar{\Pi}\,\frac{\beta\{(b^2/r^2) + 1\} - \{(b^2/r^2) - 1\}}{\beta\{(b^2/a^2) + 1\} - \{(b^2/a^2) - 1\}} \tag{16}$$

and

$$\bar{\sigma}_{\theta\theta} = \bar{\Pi}\,\frac{\beta\{(b^2/r^2) - 1\} - \{(b^2/r^2) + 1\}}{\beta\{(b^2/a^2) + 1\} - \{(b^2/a^2) - 1\}} \tag{17}$$

where $\bar{\Pi}$ is the Laplace transform of Π, which is in general a function of the time, t. Next replace the elastic constants in eqs. (16) and (17) by the corresponding p-varying moduli. Now the elastic constants of the cylinder ν and E are both contained in β as given by eq. (13). Using Table (3.1), we see that ν must be replaced by $[Y_\nu(p) - Y_s(p)]/[2Y_\nu(p) + Y_s(p)]$ and E by $3[2/Y_s(p) + 1/Y_\nu(p)]^{-1}$.

Hence, for the viscoelastic material,

$$\beta(p) = \frac{Y_\nu(p) + 2Y_s(p)}{Y_\nu(p) - Y_s(p) - (1/\alpha)Y_s(p)\{2Y_\nu(p) + Y_s(p)\}} \tag{18}$$

where

$$\alpha = \text{constant} = \frac{hE_B}{(1 - \nu_B^2)b}. \tag{19}$$

Substitute back for β into eqs. (16) and (17). Inversion of the transforms gives the solutions for σ_{rr} and $\sigma_{\theta\theta}$ in a viscoelastic cylinder.

As an example we consider a material that is incompressible and Voigt deviatoric and subject to an internal pressure

$$\Pi(t) = \Pi_0\{1 - \exp(-nt)\}H(t). \tag{20}$$

Hence $Y_\nu(p) = \infty$, $Y_s(p) = Ap + B$ where A and B are constants,

and

$$\bar{\Pi} = \Pi_0\,\frac{n}{p + n}.$$

From eq. (18)

$$\beta(p) = \left\{1 - \frac{2}{\alpha}(Ap + B)\right\}^{-1}.$$

Substituting in eqs. (16) and (17),

$$\bar{\sigma}_{rr} = -\frac{n\Pi_0}{p(p+n)}\frac{(Ap+B)\{(b^2/r^2)-1\}+\alpha}{(Ap+B)\{(b^2/a^2)-1\}+\alpha}$$

and
$$\bar{\sigma}_{\theta\theta} = \frac{n\Pi_0}{p(p+n)}.$$

Inverting the transforms,

$$\sigma_{rr} = -\Pi_0\Bigg[\frac{B\{(b^2/r^2)-1\}+\alpha}{B\{(b^2/a^2)-1\}+\alpha}$$
$$-\frac{(B-nA)\{(b^2/r^2)-1\}+\alpha}{(B-nA)\{(b^2/a^2)-1\}+\alpha}\exp(-nt)$$
$$-\frac{nA\alpha[(b^2/a^2)-(b^2/r^2)]\exp\{-t[B\{(b^2/a^2)-1\}+\alpha]/[A\{(b^2/a^2)-1\}]\}}{[B\{(b^2/a^2)-1\}+\alpha][\{nA-B\}\{(b^2/a^2)-1\}-\alpha]}\Bigg] \quad (21)$$

and

$$\sigma_{\theta\theta} = \Pi_0\Bigg[\frac{B\{(b^2/r^2)+1\}-\alpha}{B\{(b^2/a^2)-1\}+\alpha}$$
$$-\frac{(B-nA)\{(b^2/r^2)+1\}-\alpha}{(B-nA)\{(b^2/a^2)-1\}+\alpha}\exp(-nt)$$
$$+\frac{nA\alpha\{(b^2/a^2)+(b^2/r^2)\}\exp\{-t[B\{(b^2/a^2)-1\}+\alpha]/[A\{(b^2/a^2)-1\}]\}}{[B\{(b^2/a^2)-1\}+\alpha][\{nA-B\}\{(b^2/a^2)-1\}-\alpha]}\Bigg]. \quad (22)$$

For finite positive n, eq. (20) gives an internal pressure that increases continuously from 0 to Π_0. If $n \to +\infty$, $\Pi(t) \to \Pi_0 H(t)$. Hence the stresses obtaining in a sudden application of an internal pressure Π_0 are given by letting $n \to \infty$ in eqs. (21) and (22):

$$\sigma_{rr} = -\Pi_0\Bigg\{\frac{B\{(b^2/r^2)-1\}+\alpha}{B\{(b^2/a^2)-1\}+\alpha}$$
$$-\frac{\alpha\{(b^2/a^2)-(b^2/r^2)\}}{[B\{b^2/a^2\}-1\}+\alpha]\{(b^2/a^2)-1\}}$$
$$\times \exp\left[-\frac{B\{(b^2/a^2)-1\}+\alpha}{A\{(b^2/a^2)-1\}}t\right]\Bigg\} \quad (23)$$

and

$$\sigma_{\theta\theta} = \Pi_0 \left[\frac{B\{(b^2/r^2)+1\} - \alpha}{B\{(b^2/a^2)-1\} + \alpha} \right.$$

$$+ \frac{\alpha\{(b^2/a^2)+(b^2/r^2)\}}{[B\{(b^2/a^2)-1\}+\alpha]\{(b^2/a^2)-1\}}$$

$$\left. \times \exp\left[-\frac{B\{(b^2/a^2)-1\}+\alpha}{A\{(b^2/a^2)-1\}}t\right] \right] \quad (24)$$

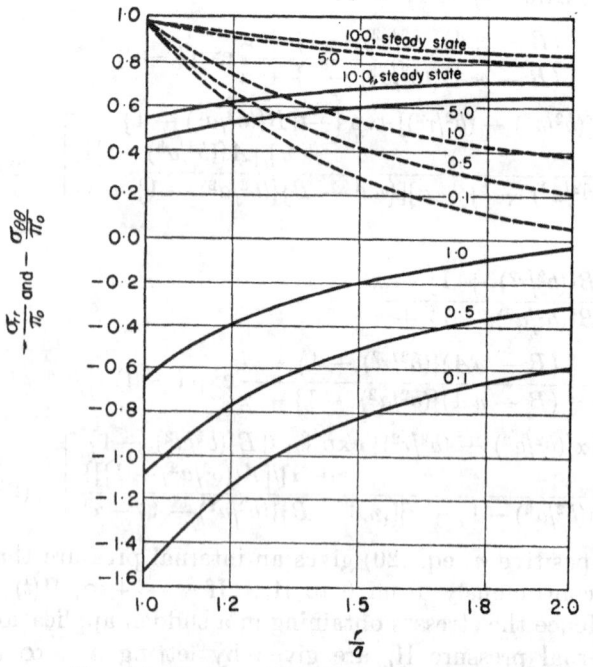

FIG. 4.1. Stresses in a reinforced incompressible Voigt cylinder, $\Pi(t) = \Pi_0 H(t)$. Numbers on curves give time in milliseconds - - - - $-\sigma_{rr}/\Pi_0$ ——— $-\sigma_{\theta\theta}/\Pi_0$.

Woodward and Radok [1] evaluated the stresses for $b/a = 2$, $A/B = 10^{-2}$ sec, $B \simeq \tfrac{2}{3} \cdot 10^5$ lb/in² and a band of steel for which $b/h = 33$, whence $\alpha \simeq 10^6$ lb/in². In Fig. (4.1), $-\sigma_{rr}/\Pi_0$ and $-\sigma_{\theta\theta}/\Pi_0$ are plotted against r/a for different values of t using eqs. (23) and (24). In Fig. (4.2), eqs. (21) and (22) are used with $n \simeq 300$ sec^{-1}. In both cases the steady state stresses are those which would obtain for an incompressible elastic cylinder with shear modulus equal to $B/2 = \frac{1}{3} \cdot 10^5 \mathrm{lb/in^2}$.

For a sudden application of pressure it is seen that the radial stress σ_{rr} is always compressive and that its absolute value rises continuously at all points. The hoop stress commences in tension but monotonically decreases algebraically at all

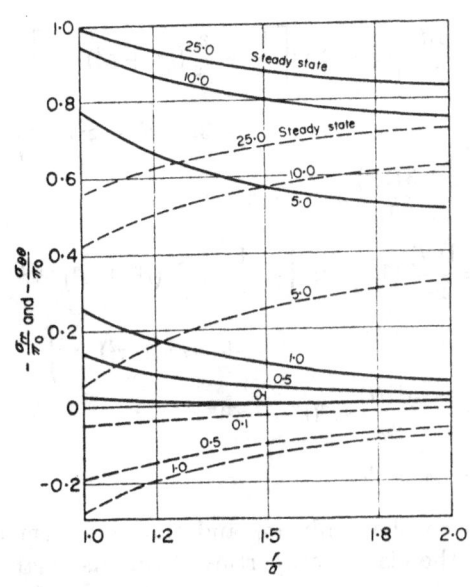

FIG. 4.2. Stresses in a reinforced incompressible Voigt cylinder $\Pi(t) = \Pi_0(1 - e^{-300t})H(t)$. Numbers on curves give time in milliseconds - - - - $-\sigma_{rr}/\Pi_0$ ——— $-\sigma_{\theta\theta}/\Pi_0$.

points. After 10 msec both stresses have nearly reached their final values. When the pressure is applied gradually the radial stress steadily increases in absolute value but the hoop stress first algebraically increases and then decreases through zero to its final negative value. Whenever times of the order of milliseconds are important, such as in rocketry, it may be of importance in design to take account of the fact that the stresses in the viscoelastic cylinder are different from those in the elastic one for about the first 20 msec.

3. Point force on a semi-infinite plane

The problem to be considered is that of a point force $\Pi(t)$ acting normally at a fixed point on a semi-infinite plane of viscoelastic material. Take cylindrical polar co-ordinates r, θ, z with origin at the point of application and the z-axis normally into the material. The stress solution for an elastic material [2] is

$$\begin{aligned}
\sigma_{rr} &= \frac{\Pi(t)}{2\pi}\left\{(1-2\nu)\left[\frac{1}{r^2} - \frac{z}{r^2}(r^2+z^2)^{-1/2}\right] \right. \\
&\qquad\qquad\left. - 3r^2z(r^2+z^2)^{-5/2}\right\}, \\
\sigma_{zz} &= -\frac{3\Pi(t)}{2\pi}z^3(r^2+z^2)^{-5/2}, \\
\sigma_{\theta\theta} &= \frac{\Pi(t)}{2\pi}(1-2\nu)\left\{-\frac{1}{r^2} + \frac{z}{r^2}(r^2+z^2)^{-1/2} \right. \\
&\qquad\qquad\left. + z(r^2+z^2)^{-3/2}\right\}, \\
\sigma_{rz} &= -\frac{3\Pi(t)}{2\pi}rz^2(r^2+z^2)^{-5/2}
\end{aligned} \quad (25)$$

and $\qquad \sigma_{r\theta} = \sigma_{z\theta} = 0.$

It can be seen that only $\sigma_{\theta\theta}$ and the first term in σ_{rr} are dependent on the elastic properties of the material—through the factor $1-2\nu$. σ_{zz}, σ_{rz}, $\sigma_{r\theta}$, $\sigma_{z\theta}$ and the second term in σ_{rr} are independent of the elastic constants and are therefore the same for a viscoelastic material. In particular if the material is incompressible, i.e. $\nu = \frac{1}{2}$, then the stress solution for a viscoelastic material is identical to that for the elastic solid; it depends only on the current value of $\Pi(t)$, not upon former values.

In general for a viscoelastic material $\sigma_{\theta\theta}$ and the first term in σ_{rr}, denoted by σ'_{rr}, must be found by the correspondence principle. Referring to Table (3.1) we see that $1-2\nu$ must be replaced by

$$1 - 2\frac{Y_\nu(p) - Y_s(p)}{2Y_\nu(p) + Y_s(p)} = \frac{3Y_s(p)}{2Y_\nu(p) + Y_s(p)}.$$

The Laplace transforms of σ'_{rr} and of $\sigma_{\theta\theta}$ for a viscoelastic material are therefore

$$\bar{\sigma}'_{rr} = \frac{1}{2\pi}\left[\frac{1}{r^2} - \frac{z}{r^2}(r^2 + z^2)^{-1/2}\right]\frac{3\overline{\Pi}(p)Y_s(p)}{2Y_\nu(p) + Y_s(p)}$$

and

$$\bar{\sigma}_{\theta\theta} = \frac{1}{2\pi}\left[-\frac{1}{r^2} + \frac{z}{r^2}(r^2 + z^2)^{-1/2} + z(r^2 + z^2)^{-3/2}\right]\frac{3\overline{\Pi}(p)Y_s(p)}{2Y_\nu(p) + Y_s(p)}.$$

Inverting these transforms enables σ'_{rr} and $\sigma_{\theta\theta}$ to be found.

Following Lee [3], let us invert the transforms for a material that is Voigt deviatoric and elastic dilatational, i.e.

$$\left.\begin{array}{l} Y_s(p) = Ap + B \\ \text{and } Y_\nu(p) = C \quad \text{where } A, B \text{ and } C \text{ are constants.} \end{array}\right\} \quad (26)$$

Then
$$\frac{3Y_s(p)}{2Y_\nu(p) + Y_s(p)} = 3\frac{Ap + B}{Ap + B + 2C}.$$

The inverse transform of $3(Ap + B)/(Ap + B + 2C)$ is

$$3\delta(t) - \frac{6C}{A}\exp\left\{\frac{-(B + 2C)t}{A}\right\}H(t).$$

By the convolution theorem the inverse transform of $3\overline{\Pi}(p)(Ap + B)/(Ap + B + 2C)$ is

$$\int_0^t \left\{3\delta(\tau) - \frac{6C}{A}\exp\left(-\frac{B + 2C}{A}\tau\right)H(t)\right\}\Pi(t - \tau)\,d\tau.$$

Therefore

$$\sigma'_{rr} = \frac{1}{2\pi}\left[\frac{1}{r^2} - \frac{z}{r^2}(r^2 + z^2)^{-1/2}\right]$$
$$\times \left\{3\Pi(t) - \frac{6C}{A}\int_0^t \exp\left(-\frac{B + 2C}{A}\tau\right)\Pi(t - \tau)\,d\tau\right\}$$

and

$$\sigma_{\theta\theta} = \frac{1}{2\pi}\left[-\frac{1}{r^2} + \frac{z}{r^2}(r^2 + z^2)^{-1/2} + z(r^2 + z^2)^{-3/2}\right]$$
$$\times \left\{3\Pi(t) - \frac{6C}{A}\int_0^t \exp\left(-\frac{B + 2C}{A}\tau\right)\Pi(t - \tau)\,d\tau\right\}.$$

In particular for $\Pi(t) = \Pi_0 H(t)$, the complete stress solution for the material whose p-varying moduli are given by equation (26) is

$$\begin{aligned}\sigma_{rr} &= \frac{3\Pi_0}{2\pi}\left\{\frac{1}{2C+B}\left[B + 2C\exp\left(-\frac{B+2C}{A}t\right)\right]\right. \\ &\quad \times \left[\frac{1}{r^2} - \frac{z}{r^2}(r^2+z^2)^{-1/2}\right] - r^2 z(r^2+z^2)^{5/2}\Big\}, \\ \sigma_{\theta\theta} &= \frac{3\Pi_0}{2\pi}\left\{\frac{1}{2C+B}\left[B + 2C\exp\left(-\frac{B+2C}{A}t\right)\right]\right. \\ &\quad \times \left[-\frac{1}{r^2} + \frac{z}{r^2}(r^2+z^2)^{-1/2} + z(r^2+z^2)^{-3/2}\right]\Big\}\end{aligned} \quad (27)$$

and the other stress equations as in eqs. (25) with $\Pi(t) = \Pi_0$.

The components of displacements u and w are given by Timoshenko [2] for an elastic solid ($v \equiv 0$). These can be treated in the same manner as the stress components to give the displacement for a viscoelastic material.

4. Moving point force on a semi-infinite plane

The stress solution in the last section has been extended by Lee [3] to the case where the point force moves along a line on the plane. He chooses this line as the x-axis and for simplicity considers only the stress component σ_{xx} at a point on the plane $y = 0$. If the point moves so that its x coordinate is given, $x = \xi(t)$, then the normal stress over the plane at any point at any time can be written

$$p(x, y, t) = \Pi(t)\,\delta(x - \xi(t))\,\delta(y). \quad (28)$$

Now the solution for an elastic material subject to this normal stress depends only upon the current stress and is therefore given by the equations of the last section. In particular σ_{xx} on $y = 0$ is equal to the previous σ_{rr} for an origin at $(\xi(t), 0, 0)$. Hence

$$(\sigma_{xx})_{y=0} = \frac{\Pi(t)}{2\pi}\left\{(1-2\nu)\left[\frac{1}{(x-\xi)^2} - \frac{z}{(x-\xi)^2}\right.\right.$$
$$\times ((x-\xi)^2 + z^2)^{-\frac{1}{2}}\bigg] - 3(x-\xi)^2 z ((x-\xi)^2 + z^2)^{-5/2}\bigg\}. \quad (29)$$

By the correspondence principle, the solution for a semi-infinite plane of a viscoelastic material subject to the normal stress of equation (28) is given by applying the Laplace transform to eq. (29), remembering that ξ is a function of t, replacing the elastic moduli by the corresponding p-varying moduli and inverting the result.

For a viscoelastic material, whose p-varying moduli are given by eq. (26), use of the convolution theorem gives the result

$$(\sigma_{xx})_{y=0} = \frac{3}{2\pi} \Pi(t) \left[\frac{1}{(x-\xi(t))^2} - \frac{z}{(x-\xi(t))^2} \{(x-\xi(t))^2 + z^2\}^{-1/2} - z(x-\xi(t))^2 \{(x-\xi(t))^2 + z^2\}^{-5/2} \right]$$
$$- \frac{3C}{\pi A} \int_0^t d\tau \exp\left(-\frac{2C+B}{A}\tau\right) \Pi(t-\tau) \left[\frac{1}{(x-\xi(t-\tau))^2} - \frac{z}{(x-\xi(t-\tau))^2} \{(x-\xi(t-\tau))^2 + z^2\}^{-1/2} \right]. \quad (30)$$

The method of this section can be used to estimate the stress and displacement produced in the material of a road, which to a first approximation is linear viscoelastic, by a car moving in a straight line across its surface. The car is replaced by four points at the vertices of a rectangle and the stresses and displacements produced by these four moving points are summed.

5. Another form of the correspondence principle

If the stress–strain relations are taken in differential form the governing equations for a quasi-static problem are

$$\epsilon_{ij} = \tfrac{1}{2}(u_{j,i} + u_{i,j}), \quad (1)$$

$$\sigma_{ij,j} + \rho X_i = 0 \quad (2)$$

$$P_v(D)\sigma_{kk} = Q_v(D)\epsilon_{kk} \quad \text{and} \quad P_s(D)s_{ij} = Q_s(D)e_{ij} \quad (31)$$

where each pair $P_v(D)$ and $Q_v(D)$, and $P_s(D)$ and $Q_s(D)$, satisfy the conditions stated in Chapter 2, Section 8. The derivatives with respect to time occur only in eqs. (31), and the derivatives with respect to the spacial co-ordinates x_i only in eqs. (1) and (2). If one of the dependent variables is differentiated (or integrated) with respect to time and space, it is immaterial in which order the differentiation (or integration) occurs. Consequently, if it is possible to solve eqs. (1), (2) and (31) for any of the dependent variables, treating $P_v(D)$, $Q_v(D)$, $P_s(D)$ and $Q_s(D)$ as constants during the solution and using the spacial boundary conditions at time t, then the resultant equation for the dependent variable will be an ordinary differential equation in the time variable t. All the space differentiations and integrations have been carried out first, leaving the time differentiations and integrations until the end.

For an elastic solid the governing equations are identical to (1), (2) and (31), except that $Q_v(D)/P_v(D)$ is replaced by $3k$ and $Q_s(D)/P_s(D)$ by 2μ. The elastic solution is therefore the same as the spacial part of the viscoelastic solution, since no differentiations with respect to time occur in the governing equations for the elastic problem. Hence the second form of the correspondence principle for quasi-static problems:

"If a particular quasi-static problem can be solved for one of the dependent variables for an elastic solid at any time t, then that variable in the viscoelastic solution satisfies the ordinary differential equation given by replacing the elastic constants in the elastic solution by the appropriate viscoelastic operators."

The appropriate operators are given by Table (3.1) with Y_v replaced by $Q_v(D)/P_v(D)$ and Y_s by $Q_s(D)/P_s(D)$.

This form of the principle is more general than that given previously as it enables the position of the boundaries to be functions of time. Three illustrations of its use will be given—in the next two sections and immediately below to solve again the problem of Section 4.

The elastic solution in Section 4 is given by eq. (29). For the viscoelastic material whose p-varying moduli are given by eqs. (26), $1 - 2\nu$ must be replaced by

$$\frac{3Y_s(D)}{2Y_v(D) + Y_s(D)} = 3\frac{AD + B}{AD + B + 2C}.$$

Hence for the viscoelastic material $(\sigma_{xx})_{y=0}$ is given by

$$(\sigma_{xx})_{y=0} = \frac{3}{2\pi} \frac{AD+B}{AD+B+2C} \left\{ \Pi(t) \left[\frac{1}{(x-\xi(t))^2} \right. \right.$$
$$\left. \left. - \frac{z}{(x-\xi(t))^2} ((x-\xi(t))^2 + z^2)^{-1/2} \right] \right\}$$
$$- \frac{3\Pi(t)}{2\pi} (x-\xi(t))^2 z((x-\xi(t))^2 + z^2)^{-5/2},$$

and eq. (30) is immediately derived from the theory of linear operators.

6. Indentation of an incompressible semi-infinite plane by a smooth rigid sphere

This solution is due to Lee and Radok [4]. In this problem the boundary conditions on the surface of the plane alter as the sphere indents the plane. Where the sphere is in contact with the plane, the conditions are no tangential stress and given normal displacement; where the sphere is not in contact, the conditions are no tangential stress and no normal stress. The correspondence principle as stated in Section 1 cannot be used because the boundaries are moving.

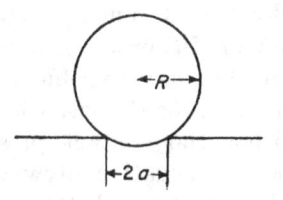

This difficulty does not arise with the solution for an elastic plane because that depends only upon the current configuration. The relation between the radius of contact, $a(t)$, and the total normal force between sphere and plane, $\Pi(t)$, is given by

$$\Pi(t) = \frac{16\mu}{3R} \{a(t)\}^3, \tag{32}$$

where R is the radius of the sphere and μ is the shear modulus of the elastic material of the plane, assumed incompressible. This result can be deduced from the theory of contact of two elastic spheres [5].

Now consider that a smooth rigid sphere is placed on a semi-infinite viscoelastic plane so that its radius of contact gradually increases with time. At each time the corresponding elastic problem has the same radius of contact and the boundary conditions are identical in the two cases. The corresponding elastic problem is different at each instant of time. The governing differential equations for the two problems differ only in that the deviatoric stress–strain relation in the elastic case is

$$s_{ij} = 2\mu e_{ij},$$

whereas in the viscoelastic case it is given by eqs. (31). Hence, by the correspondence principle as stated in Section 5 and by Table (3.1) for a viscoelastic material

$$P_s(D)\Pi(t) = \frac{8}{3R} Q_s(D)\{a(t)\}^3. \tag{33}$$

Equation (33) provides one equation between $\Pi(t)$ and $a(t)$. The other equation will be part of the boundary conditions. However it is not possible to impose any form whatsoever on one of $\Pi(t)$ or $a(t)$ if use is to be made of eq. (33). A solution for the elastic problem always exists but the corresponding solution for the viscoelastic problem is not correct if it causes either the contact pressure between sphere and plane to become negative anywhere, or the normal displacement of the plane to be such that it would enter the space occupied by the sphere at any point.

We consider the case in which the force between sphere and plane is due solely to the weight Π_0 of the sphere. If the sphere was placed on the plane at time $t = 0$,

$$\Pi(t) = \Pi_0 H(t). \tag{34}$$

Apply the one-sided Laplace transform to eqs. (33) and (34):

$$P_s(p)\overline{\Pi} = \frac{8}{3R} Q_s(p)\overline{a^3} \tag{35}$$

and

$$\overline{\Pi} = \Pi_0/p. \tag{36}$$

Eliminating $\overline{\Pi}$ from eqs. (35) and (36) and using eq. (5),

$$\overline{a^3} = \frac{3R\Pi_0}{8} \frac{P_s(p)}{pQ_s(p)} = \frac{3R\Pi_0}{8} \frac{J_s(p)}{p}.$$

But $J_s(p)/p$ is the Laplace transform of the strain response to a stress $H(t)$. Therefore

$$a^3(t) = \frac{3R\Pi_0}{8}\left(\frac{1}{E_s} + \frac{1}{\eta_s}t + \psi_s(t)\right), \qquad (37)$$

where E_s, η_s and $\psi_s(t)$ are the instantaneous elastic modulus, the long term viscous flow constant and the creep functions for deviatoric stress respectively. In particular for a Maxwell incompressible plane, $(D/E + 1/\eta)s_{ij} = e_{ij}$, the sphere penetrates into the plane to the elastic depth immediately and then $a^3(t)$ continues to increase at a constant rate; for a Voigt incompressible plane, $s_{ij} = (\eta D + E)e_{ij}$, the penetration is initially zero but it asymptotically approaches the elastic value for large values of the time. Equation (37) may provide a method of determining the creep function for an incompressible material experimentally if the penetration is small over a long period of time.

7. Biot's stability problem

Biot [6] considers the stability of a viscoelastic plate of constant thickness sandwiched in an infinitely extended medium of another viscoelastic material and subject to a compressive normal stress Π acting parallel to the plate. The direction of action of Π is chosen as the x axis and the normal to the plate as the z axis.

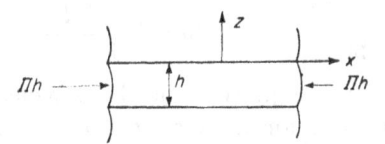

There is assumed to be no friction between plate and medium so that the interaction pressure between them is normal to their common surface. Let the total force per unit area exerted by the medium on the plate be q and let w be the displacement of the plate, both parallel to the z-axis. Then for an elastic plate

$$\frac{Eh^3}{12(1-\nu^2)}\frac{d^4w}{dx^4} + \Pi h \frac{d^2w}{dx^2} = q, \qquad (38)$$

where E and ν are the Young's modulus and Poisson's ratio of the plate. q and w are also related by the equations of deformation of the medium. Assuming the deflection w to be sinusoidal in x,

i.e.
$$w = w_0 \cos lx, \tag{39}$$

the normal deflection of a semi-infinite plane subject to such a surface loading is also sinusoidal of the form

$$q' = q_0 \cos lx.$$

If the medium is elastic with constants E' and ν', then (see e.g. [7])

$$w = 2q' \frac{1 - \nu_1^2}{E_1 l}. \tag{40}$$

Now q' is the force exerted per unit area by the medium on one side of the layer, hence

$$q = -2q', \tag{41}$$

since the layer is surrounded by the medium on both sides. (If the layer were in contact with the medium on one side only, the other side being stress free, then we would have $q = -q'$.)

Substituting from eqs. (40) and (41) into eq. (38),

$$\frac{Eh^3}{12(1 - \nu^2)} \frac{d^4w}{dx^4} + \Pi h \frac{d^2w}{dx^2} + \frac{E_1 l}{1 - \nu_1^2} w = 0. \tag{42}$$

Using eq. (39),

$$\frac{Eh^3 l^4}{12(1 - \nu^2)} - \Pi h l^2 + \frac{E_1 l}{1 - \nu_1^2} = 0 \tag{43}$$

is the characteristic equation for the wavelength $L\ (= 2\pi/l)$ of the deformation for an elastic plate surrounded by an elastic medium. Equation (43) is a cubic equation in lh. The least positive value of Π giving a positive real root lh is

$$\Pi_{cr} = \frac{3}{2} \frac{E_1}{1 - \nu_1^2} \left(\frac{E(1 - \nu_1^2)}{6E_1(1 - \nu^2)} \right)^{1/3}, \tag{44}$$

for which

$$l_{cr} h = \left(\frac{6E_1(1 - \nu^2)}{E(1 - \nu_1^2)} \right)^{1/3}; \tag{45}$$

hence
$$\Pi_{cr} = \frac{3}{2} \frac{E_1}{(1 - \nu_1^2) l_{cr} h}. \tag{46}$$

QUASI-STATIC PROBLEMS

For a viscoelastic plate in a viscoelastic medium, by the correspondence principle and Table (3.1), $E/(1 - \nu^2)$ must be replaced by $B(D)$, where

$$B(D) = \frac{\{2Y_\nu(D) + Y_s(D)\}Y_s(D)}{Y_\nu(D) + 2Y_s(D)}, \qquad (47)$$

and $E_1/(1 - \nu_1^2)$ by $B_1(D)$, where

$$B_1(D) = \frac{\{(2Y_{\nu 1}(D) + Y_{s1}(D)\}Y_{s1}(D)}{Y_{\nu 1}(D) + 2Y_{s1}(D)}. \qquad (48)$$

The characteristic eq. (43) becomes

$$\tfrac{1}{12}B(D)l^3h^3 - \Pi lh + B_1(D) = 0. \qquad (49)$$

This equation can be considered as a relation between Π, l and the time parameter, D. Positive values of D correspond to deflexions containing a time factor e^{Dt}, i.e. an equation identical to eq. (49) could be derived if we assumed $w = f(x)e^{Dt}$ where D is a positive constant; hence D is a measure of the rate of growth of the deformation. Starting from unstable equilibrium, one would expect that configuration to be observed which had the greatest rate of growth; the corresponding wavelength is known as the "dominant" wavelength, L_d. It follows from eqs. (47) (48) and (2.86) that $B(D)$ and $B_1(D)$ are monotonically increasing functions of D. For a fixed value of Π, the greatest value of D giving a positive real root lh of eq. (49) is given by

$$\{B_1(D)\}^2 B(D) = \tfrac{16}{9}\Pi^3. \qquad (50)$$

The corresponding value of the wavelength is $L_d(= 2\pi/l_d)$,

where
$$l_d h = \left(\frac{6B_1(D)}{B(D)}\right)^{1/3}, \qquad (51)$$

or
$$L_d = 2\pi h \left(\frac{B(D)}{6B_1(D)}\right)^{1/3}. \qquad (52)$$

As in the elastic case, the theory is only valid if $L_d \gg h$.

From eq. (50) if $B_1(0) \neq 0$, if $B(0) \neq 0$, and if $\Pi_0^3 = \tfrac{9}{16}(B_1(0))^2 B(0)$, the plate is stable for any $\Pi < \Pi_0$. For

$\Pi = \Pi_0$ the plate is unstable but the rate of deformation is zero. If $B_1(\infty)$ and $B(\infty)$ are finite and if $\Pi_\infty{}^3 = (9/16)\{B_1(\infty)\}^2 B(\infty)$, then for any $\Pi \geqslant \Pi_\infty$ the rate of deformation is infinite, i.e. the deformation is elastic. If $B(D) = CB_1(D)$ where C is a constant, then the dominant wavelength is independent of the pressure and the plate always deforms in the same manner. However, if $B(D) \neq CB_1(D)$ for any constant C, then the mode of deformation depends on the pressure; if $\Pi_0 < \Pi_1 < \Pi_2 < \Pi_\infty$ and if pressures Π_1 and Π_2 are applied successively, then the resulting deformation is the sum of two sine waves of wavelengths given by eqs. (50) and (52) with $\Pi = \Pi_1$ and with $\Pi = \Pi_2$ respectively. Examples of the deformation that occurs with particular types of viscoelastic materials, both in plate and medium, are given by Biot [6]. The reader is referred to his paper. The results may have applications to plastics, to sandwich panels, to metals at high temperatures, and to the folding of stratified geological formations. Biot has recently extended the analysis to include the presence of a gravitational field acting normal to the layer [8].

REFERENCES

1. W. B. WOODWARD and J. R. M. RADOK: *Brown University Technical Report PA-TR-14* (1955).
2. S. TIMOSHENKO: *Theory of Elasticity* (Section 105, McGraw-Hill, 1934).
3. E. H. LEE: *Quart. Appl. Math.* **13** (1955) 183.
4. E. H. LEE and J. R. M. RADOK: *Proc. 9th Int. Congr. Appl. Mech.*, Brussels, 1956.
5. S. TIMOSHENKO and J. N. GOODIER: *Theory of Elasticity*, Second Edition (Section 126, McGraw-Hill, 1957).
6. M. A. BIOT: *Proc. Roy. Soc.* A **242** (1958) 444.
7. M. A. BIOT: 1937, *Trans. Amer. Soc. Mech. Engrs.* **59**, A1–A7.
8. M. A. BIOT: *J. Franklin Inst.*, **267**, (1959), 211.

CHAPTER 5

STRESS ANALYSIS III: DYNAMIC PROBLEMS

1. The correspondence principle

In this chapter, as in Chapter 4, we are concerned with the determination of stress and/or displacement in a viscoelastic body initially dead (i.e. $\sigma_{ij} \equiv 0$, $u_i \equiv 0$ for $t < 0$), but now the inertia terms are included in the equations of motion. The governing differential equations are taken in the form

$$\epsilon_{ij} = \tfrac{1}{2}(u_{j,i} + u_{i,j}), \tag{1}$$

$$\sigma_{ij,j} + \rho X_i = \rho \frac{\partial^2 u_i}{\partial t^2}, \tag{2}$$

$$P_v(D)\sigma_{kk} = Q_v(D)\epsilon_{kk} \quad \text{and} \quad P_s(D)s_{ij} = Q_s(D)e_{ij}. \tag{3}$$

It is seen that time derivatives occur in eqs. (2) and therefore the correspondence principle in the form stated for quasi-static problems in Section 5 of the last chapter is no longer applicable. In problems of impact a Fourier time analysis of the applied forces generally shows a large contribution from the components of high frequency or of small characteristic time. Since the experimentally measurable properties of viscoelastic materials under such conditions are the complex moduli or compliances, a method of solving eqs. (1) to (3) is required which expresses the material properties in that form. It is provided by the one-sided Fourier transform.

The one-sided Fourier transform of $f(t)$ is defined by†

$$\bar{f}(\omega) = \int_0^\infty \exp(-i\omega t) f(t)\, dt. \tag{4}$$

The inverse transform is

$$f(t) = \frac{1}{\pi} R\left[\int_0^\infty \exp(i\omega t) \bar{f}(\omega)\, d\omega\right]. \tag{5}$$

† In this chapter a bar denotes a one-sided Fourier transform.

If $f(t) = 0$ for $t < 0$, then

$$\int_0^\infty \exp(-i\omega t) \frac{df(t)}{dt} dt = [f(t) \exp(-i\omega t)]_0^\infty$$
$$+ i\omega \int_0^\infty \exp(-i\omega t) f(t) dt = i\omega \bar{f}(\omega),$$

provided, as is assumed throughout this chapter, $f(t)\exp(-i\omega t)$ converges to zero as $t \to \infty$. By induction it can be shown that the transform of $d^n f(t)/dt^n$ is $(i\omega)^n \bar{f}(\omega)$. It follows that the transform of $\sum_{r=0}^n a_r d^r f(t)/dt^r$ is $\sum_{r=0}^n a_r (i\omega)^r \bar{f}(\omega)$. Hence, if the transform is applied to eqs. (3),

$$P_v(i\omega)\bar{\sigma}_{kk} = Q_v(i\omega)\bar{\epsilon}_{kk} \quad \text{and} \quad P_s(i\omega)\bar{s}_{ij} = Q_s(i\omega)\bar{e}_{ij}.$$

But $Q_v(i\omega)/P_v(i\omega) = Y_v(i\omega)$ and $Q_s(i\omega)/P_s(i\omega) = Y_s(i\omega)$, where Y_v and Y_s are the complex moduli; therefore

$$\bar{\sigma}_{kk} = Y_v(i\omega)\bar{\epsilon}_{kk} \quad \text{and} \quad \bar{s}_{ij} = Y_s(i\omega)\bar{e}_{ij}. \tag{6}$$

The use of the one-sided Fourier transform has expressed the stress–strain equation in the desired form.

If the transform is applied to eqs. (1) and (2)

$$\bar{\epsilon}_{ij} = \tfrac{1}{2}(\bar{u}_{j,i} + \bar{u}_{i,j}) \tag{7}$$

and
$$\bar{\sigma}_{ij,j} + \rho \bar{X}_i + \rho \omega^2 \bar{u}_i = 0. \tag{8}$$

The governing equations for an elastic solid differ from eqs. (1), (2) and (3) only in that eqs. (3) are replaced by $\sigma_{kk} = 3k\epsilon_{kk}$ and $s_{ij} = 2\mu e_{ij}$. Their transforms differ only in that $Y_v(i\omega)$ and $Y_s(i\omega)$ are replaced by $3k$ and 2μ respectively. Therefore a correspondence principle for dynamic problems can be stated:

"The solution for a dynamic problem for a viscoelastic material can be obtained from the solution for the corresponding problem for an elastic solid by applying the one-sided Fourier transform to the elastic solution, replacing the elastic constants by the corresponding viscoelastic complex moduli (or compliances) and finally inverting the transform."

It is understood that the boundary conditions for the two problems are identical and that in each problem the acceleration term can be expressed as $\partial^2 u_i / \partial t^2$.

At first sight it may appear that this principle enables a large number of dynamic problems to be solved in viscoelasticity. Unfortunately this is not the case because not many dynamic problems in elasticity have been solved. (Sinusoidal oscillation problems are excluded from consideration by the condition that the material is initially dead.) This is in contradistinction to the situation for quasi-static problems in viscoelasticity where the corresponding problem in elasticity is static and many particular solutions are known. The dynamic problem most studied to date is that of the propagation of longitudinal waves down a rod and this will be treated in the next section both with and without the correspondence principle. The correspondence principle will then be used to consider radially symmetric impact on a spherical cavity and finally radially symmetric impact on a plate.

2. The propagation of longitudinal waves along semi-infinite rods

The one-dimensional theory of wave propagation in rods will be used. The theory is satisfactory provided the wavelength of the disturbance is large compared to a typical diameter of the rod.† The same theory, with only a change of moduli, describes the propagation of plane dilatational waves in a direction normal to the plane surface of a semi-infinite medium. In the latter case there is no limitation on wavelength.

Co-ordinate axes are chosen so that the x axis lies along the rod with the origin at the end of the rod. The theory assumes that, with respect to these axes, all components of stress are zero except the normal component parallel to the x axis, denoted by σ. Let ϵ be the normal component of strain and u the component of displacement parallel to the x axis.

The governing equations for a viscoelastic material are

$$\frac{\partial \sigma}{\partial x} = \rho \frac{\partial^2 u}{\partial t^2}, \epsilon = \frac{\partial u}{\partial x} \quad \text{and} \quad P(D)\sigma = Q(D)\epsilon. \qquad (9)$$

† e.g. H. Kolsky in *Stress Waves in Solids* shows that the theory is satisfactory for bars of circular section if the radius wave-length ratio is less than 0·1 [1].

From Table (3.1), since the corresponding modulus for the elastic rod is Young's modulus,

$$\frac{Q(D)}{P(D)} = 3\left(\frac{2P_s(D)}{Q_s(D)} + \frac{P_v(D)}{Q_v(D)}\right)^{-1}$$

$$= \frac{3Q_s(D)Q_v(D)}{2P_s(D)Q_v(D) + P_v(D)Q_s(D)}.$$

Eliminating σ and ϵ,

$$\rho D^2 P(D)u = Q(D)\frac{\partial^2 u}{\partial x^2}. \tag{10}$$

Let $P(D)$ be of order n in D. It follows from Chapter 2, Section 8, that the orders of $Q_s(D)$ and $Q_v(D)$ are either equal to or one greater than the orders of $P_s(D)$ and $P_v(D)$ respectively. If both orders are one greater, then $Q(D)$ is of order $n+1$ and the highest order derivative occurring in eq. (10) is $\partial^{n+3}u/\partial t^{n+1}\partial x^2$. In this case the lines $t =$ constant (twice) and $x =$ constant ($n+1$ times) and the characteristics of equation (10) and therefore a disturbance at any point on the rod has instantaneous effects at all points of the rod. Otherwise the orders of $Q(D)$ and $P(D)$ are equal and the highest order derivatives occurring in eq. (10) are

$$q_n \frac{\partial^{n+2}u}{\partial t^n \partial x^2} - \rho p_n \frac{\partial^{n+2}u}{\partial t^{n+2}},$$

where q_n and p_n are the coefficients of the highest order terms in P and Q. The lines $x \pm \sqrt{(q_n/\rho p_n)}\, t =$ constant (each once) and $x =$ constant (n times) are the characteristics of the equation. A disturbance at a point on the rod has no effect at a distance L until a time $\sqrt{(\rho p_n/q_n)}\, L$ has elapsed. Equality or non-equality of orders of $P(D)$ and $Q(D)$ corresponds to whether the material does or does not exhibit instantaneous elasticity in simple tension. If all actual materials exhibit some instantaneous elasticity, then instantaneous transmission of a disturbance is impossible in practice. To investigate the solution with instantaneous elasticity in more detail, the correspondence principle will be employed.

For an elastic rod eq. (10) becomes

$$\rho \frac{\partial^2 u}{\partial t^2} = E \frac{\partial^2 u}{\partial x^2} \tag{11}$$

with solution
$$u = f(x - ct) + g(x + ct)$$
where $f(x)$ and $g(x)$ are arbitrary functions of x and
$$c^2 = \frac{E}{\rho}. \tag{12}$$

For a uniform semi-infinite rod subject only to force or prescribed displacement at the end $x = 0$, $g(x + ct) \equiv 0$ and

Whence
$$\left.\begin{array}{c} u = f(x - ct). \\ v = \dfrac{\partial u}{\partial t} = -cf'(x - ct), \\ \epsilon = \dfrac{\partial u}{\partial x} = f'(x - ct), \\ \sigma = E\epsilon = Ef'(x - ct). \end{array}\right\} \tag{13}$$

and

The form of f is determined by the conditions on the boundary $x = 0$. We shall consider the effect of an impulsive pressure.†

If the impulse is of magnitude I and is applied to the end $x = 0$ at time $t = 0$, the boundary condition is
$$\sigma_{x=0} = -I\,\delta(t).$$
Substituting from eq. (13) for an elastic rod,
$$Ef'(-ct) = -I\,\delta(t).$$
Integrating with respect to t,
$$u = f(x - ct) = \frac{cI}{E} H\!\left(t - \frac{x}{c}\right), \quad v = \frac{cI}{E}\delta\!\left(t - \frac{x}{c}\right),$$
$$\epsilon = -\frac{I}{E}\delta\!\left(t - \frac{x}{c}\right) \quad \text{and} \quad \sigma = -I\,\delta\!\left(t - \frac{x}{c}\right) \tag{14}$$

Let us find the stress for a viscoelastic rod. Applying the one-sided Fourier transform to the stress in eq. (14),
$$\bar{\sigma} = -I\int_0^\infty \exp(-i\omega t)\,\delta\!\left(t - \frac{x}{c}\right) dt = -I\exp(-i\omega x/c).$$

† The effect of a constant velocity applied at time $t = 0$ has been considered by Lee and Kanter [2], Morrison [3], and Lee and Morrison [4].

Replacing c by $\{\rho J(i\omega)\}^{-1/2}$ the transform of the stress for the viscoelastic rod is

$$\bar{\sigma} = -I \exp[-i\omega x \sqrt{\{\rho J(i\omega)\}}];$$

and the stress itself

$$\sigma = -\frac{I}{\pi} R\left[\int_0^\infty \exp\{i\omega[t - \sqrt{\{\rho J(i\omega)\}}x]\}\, d\omega\right]. \quad (15)$$

For a Maxwell material $J(i\omega) = 1/E + 1/i\omega\eta$. After substitution in eq. (15),

$$\sigma = -I \exp\{-(E/2\eta)t\}\left[\delta\left(t - \sqrt{\frac{\rho}{E}}x\right) + \frac{\sqrt{\rho E}x}{2\eta}\right.$$
$$\left.\times \left\{I_1\left(\frac{E}{2\eta}\sqrt{t^2 - \frac{\rho}{E}x^2}\right)\right\}\left(t^2 - \frac{\rho}{E}x^2\right)^{-1/2} H\left(t - \sqrt{\frac{\rho}{E}}x\right)\right],$$
$$(16)$$

where I_1 is the Bessel function of imaginary argument of the first order. The impulse is propagated with velocity $\sqrt{(E/\rho)}$ and attenuation $\sqrt{(\rho E/2\eta)}$ and is followed by a stress wave of finite amplitude.

For the general viscoelastic material exhibiting instantaneous elasticity, the asymptotic expansion† of $\sqrt{\{\rho J(i\omega)\}}$ is required.

$$J(i\omega) = \frac{p_n(i\omega)^n + p_{n-1}(i\omega)^{n-1} + \ldots + p_0}{q_n(i\omega)^n + q_{n-1}(i\omega)^{n-1} + \ldots + q_0}.$$

Hence

$$J(i\omega) \sim \frac{p_n}{q_n}\bigg\{1 - \frac{i}{\omega}\left(\frac{p_{n-1}}{p_n} - \frac{q_{n-1}}{q_n}\right)$$
$$+ \frac{1}{\omega^2}\left(-\frac{p_{n-2}}{p_n} + \frac{q_{n-2}}{q_n} - \frac{q_{n-1}^2}{q_n^2} + \frac{p_{n-1}q_{n-1}}{p_n q_n}\right) + O\left(\frac{1}{\omega^3}\right)\bigg\}$$

and

$$\sqrt{\{\rho J(i\omega)\}} \sim \sqrt{\left\{\rho\frac{p_n}{q_n}\right\}}\bigg\{1 - \frac{i}{2\omega}\left(\frac{p_{n-1}}{p_n} - \frac{q_{n-1}}{q_n}\right) + \frac{1}{8\omega^2}$$
$$\times \left(\frac{p_{n-1}^2}{p_n^2} - \frac{3q_{n-1}^2}{q_n^2} + 2\frac{p_{n-1}}{p_n}\frac{q_{n-1}}{q_n} - 4\frac{p_{n-2}}{p_n} + 4\frac{q_{n-2}}{q_n}\right) + O\left(\frac{1}{\omega^3}\right)\bigg\}.$$

† Following *Modern Analysis* by Whittaker and Watson, the symbol "$\sim$" is used to denote an asymptotic expansion [5].

Therefore $\quad \sqrt{\{\rho J(i\omega)\}} \sim \alpha - \dfrac{i\beta}{\omega} + \dfrac{\gamma}{\omega^2} + O\left(\dfrac{1}{\omega^3}\right),$ (17)

where, using $J(i\omega) = J_1(\omega) - iJ_2(\omega)$,

$$\alpha = \sqrt{\left\{\rho \dfrac{p_n}{q_n}\right\}} = \sqrt{\left\{\rho \lim_{\omega \to \infty} J_1\right\}}, \quad (18)$$

$$\beta = \dfrac{1}{2}\sqrt{\left(\rho \dfrac{p_n}{q_n}\right)\left(\dfrac{p_{n-1}}{p_n} - \dfrac{q_{n-1}}{q_n}\right)} = \dfrac{\rho}{2\alpha}\lim_{\omega \to \infty} \omega J_2 \quad (19)$$

and $\quad \gamma = \dfrac{\rho}{2\alpha}\lim_{\omega \to \infty}\omega^2\left(J_1 - \dfrac{p_n}{q_n}\right) + \dfrac{\beta^2}{2\alpha}.$ (20)

For a material exhibiting instantaneous elasticity, the use of the theory of characteristics showed that the wave front created by any disturbance travelled at velocity $1/\alpha$. Since the stress applied at $x = 0$ is a delta function, a solution is sought of the form

$$\sigma(x,t) = -Ig(x)\,\delta(t - \alpha x) - If(x,t)H(t - \alpha x). \quad (21)$$

Now

$$\dfrac{1}{\pi} R\left[\int_0^\infty \exp(i\omega t)\exp(-i\omega\alpha x - \beta x)\,d\omega\right]$$
$$= \exp(-\beta x)\delta(t - \alpha x) \quad (22)$$

and, therefore, if it can be shown that

$$f(x,t)H(t-\alpha x) = \dfrac{1}{\pi} R\left[\int_0^\infty \exp(i\omega t)\{\exp[-i\omega\sqrt{\{\rho J(i\omega)\}}x]\right.$$
$$\left. - \exp(-i\omega\alpha x - \beta x)\}\,d\omega\right] \quad (23)$$

converges for all finite x and t, then $\sigma(x,t)$ is of the form of eq. (21) with $g(x) = \exp(-\beta x)$ and $f(x,t)$ finite for all finite x and t which satisfy $t - \alpha x \geqslant 0$.

To prove the convergence of the integral in eq. (23) is equivalent to proving that, by choosing Λ sufficiently large, the integral

$$I_\Lambda = \int_\Lambda^\infty \exp(i\omega t)\{\exp[-i\omega\sqrt{\{\rho J(i\omega)\}}x]$$
$$- \exp(-i\omega\alpha x - \beta x)\}\,d\omega$$

can be made arbitrarily small for all positive finite x and t. Using eq. (17),

$$\exp\left[-i\omega\sqrt{\{\rho J(i\omega)\}}x\right] \sim \exp\left[-i\omega x\left\{\alpha + \frac{\beta}{i\omega} + \frac{\gamma}{\omega^2} + O\left(\frac{1}{\omega^3}\right)\right\}\right]$$

$$\sim \exp\left[-i\omega\alpha x - \beta x\right]\exp\left[\frac{\gamma x}{i\omega} + O\left(\frac{1}{\omega^2}\right)\right]$$

$$\sim \exp\left[-i\omega\alpha x - \beta x\right]\left\{1 + \frac{\gamma x}{i\omega} + O\left(\frac{1}{\omega^2}\right)\right\}.$$

$$I_\Lambda \sim \int_\Lambda^\infty \exp\left\{i\omega(t - \alpha x)\right\}\exp\left(-\beta x\right)\left\{\frac{\gamma x}{i\omega} + O\left(\frac{1}{\omega^2}\right)\right\}d\omega.$$

But $\int_{a>0}^\infty d\omega[\exp\{i\omega(t-\alpha x)\}]/\omega$ is convergent for all real $t - \alpha x$ and $\int_{a>0}^\infty d\omega[\exp\{i\omega(t-\alpha x)\}]/\omega^2$ is absolutely convergent for all real $t - \alpha x$ and therefore I_Λ can be made arbitrarily small by increasing Λ.

The expansion of eq. (17) can be continued to give

$$\sqrt{\{\rho J(i\omega)\}} \sim \alpha - \frac{i\beta}{\omega} + \frac{\gamma}{\omega^2} - \frac{i\lambda}{\omega^3} + \frac{\epsilon}{\omega^4} + \cdots$$

$$\exp\left[-i\omega\sqrt{\{\rho J(i\omega)\}}x\right]$$

$$\sim \exp\left(-i\omega\alpha x - \beta x\right)\exp\left\{\left(\frac{\gamma x}{i\omega} - \frac{\lambda x}{\omega^2}\right) + O\left(\frac{1}{\omega^3}\right)\right\}$$

$$\sim \exp\left(-i\omega\alpha x - \beta x\right)\left\{1 + \frac{\gamma x}{i\omega} - \left(\lambda x + \frac{\gamma^2 x^2}{2}\right)\frac{1}{\omega^2} + O\left(\frac{1}{\omega^3}\right)\right\}.$$

Therefore the integrand of eq. (23),

$$\exp\left(i\omega t\right)\exp\left[-i\omega\sqrt{\{\rho J(i\omega)\}}x\right] - \exp\left(-i\omega\alpha x - \beta x\right)$$

$$\sim \exp\left(-\beta x\right)\exp\left\{i\omega(t - \alpha x)\right\}\left\{\frac{\gamma x}{i\omega} - \left(\lambda x + \frac{\gamma^2 x^2}{2}\right)\frac{1}{\omega^2} + O\left(\frac{1}{\omega^3}\right)\right\}.$$

When the exponential factor in the integrand involving ω is $\exp(i\omega t)$, the integral, using the asymptotic expansion of the integrand for large values of ω, gives the solution for small values of t†. In the case considered here, when the factor is $\exp\{i\omega(t-\alpha x)\}$ the asymptotic expansion for larger values of ω gives the solution for small values of $t-\alpha x$, i.e. it gives the solution near the wavefront. Therefore

$$f(x,t) = \exp(-\beta x)\{\gamma x + \left(\lambda x + \frac{\gamma^2 x^2}{2}\right)(t-\alpha x) + O(t-\alpha x)^2\} \tag{24}$$

near the wavefront. If $J(i\omega)$ has been determined experimentally, the usefulness of eq. (24) is limited by the number of terms in the asymptotic expansion of $J(i\omega)$ which can be computed accurately.

To sum up, it has been shown that an impulse on the end of a semi-infinite viscoelastic rod, which exhibits instantaneous elasticity in tension, is transmitted down the rod with velocity $1/\alpha$ and attenuation β, given by eqs. (18) and (19). The attenuated impulse is followed immediately by a wave of finite amplitude whose magnitude at its head is $I\gamma x \exp(-\beta x)$. The form of wave is given by eq. (23). The integral in this last equation is convergent as $\omega \to \infty$; in general numerical evaluation is necessary.

For a material not exhibiting instantaneous elasticity,

$$J(i\omega) = \frac{p_n(i\omega)^n + p_{n-1}(i\omega)^{n-1} + \ldots + p_0}{q_{n+1}(i\omega)^{n+1} + q_n(i\omega)^n + \ldots + q_0}$$

$$\sim \frac{p_n}{i\omega q_{n+1}}\left\{1 - \frac{i}{\omega}\left(\frac{p_{n-1}}{p_n} - \frac{q_n}{q_{n+1}}\right)\right.$$

$$\left. + \frac{1}{\omega^2}\left(-\frac{p_{n-2}}{p_n} + \frac{q_{n-1}}{q_{n+1}} - \frac{q_n^2}{q^2_{n+1}} + \frac{p_{n-1}q_n}{p_n q_{n+1}}\right) + O\left(\frac{1}{\omega^3}\right)\right\}.$$

Hence

$$\sqrt{\rho J(i\omega)} \sim \sqrt{\frac{\rho p_n}{i\omega q_{n+1}}}\left\{1 - \frac{i}{2\omega}\left(\frac{p_{n-1}}{p_n} - \frac{q_n}{q_{n+1}}\right) + O\left(\frac{1}{\omega^2}\right)\right\}$$

and

$$\exp[-i\omega x\sqrt{\{\rho J(i\omega)\}}] \sim \exp\left[-x\sqrt{\left(\frac{\rho p_n i\omega}{q_{n+1}}\right)}\right.$$

$$\left. \times \left\{1 - \frac{i}{2\omega}\left(\frac{p_{n-1}}{p_n} - \frac{q_n}{q_{n+1}}\right) + O\left(\frac{1}{\omega^2}\right)\right\}\right].$$

† For a proof of this result see, e.g. [6].

Provided the integral converges, the value of $\sigma(x, t)$ for small t corresponds to the value of the inverse transform for large ω. Therefore for small t and $x \neq 0$

$$\sigma(x, t) = -\frac{I}{\pi} R\left[\int_0^\infty \exp(i\omega t) \exp\left\{-x\sqrt{\left(\frac{\rho p_n i\omega}{q_{n+1}}\right)}\right\}\right.$$
$$\left. \times \left\{1 - \frac{x}{2\sqrt{i\omega}}\sqrt{\frac{\rho p_n}{q_{n+1}}}\left(\frac{p_{n-1}}{p_n} - \frac{q_n}{q_{n+1}}\right) + O\left(\frac{1}{\omega}\right)\right\} dw\right]$$

$$= -\frac{Ix}{2}\sqrt{\frac{\rho p_n}{q_{n+1}\pi}} \exp\left(-\frac{\rho p_n}{q_{n+1}}\frac{x^2}{4t}\right)$$
$$\times \left\{t^{-3/2} - \left(\frac{p_{n-1}}{p_n} - \frac{q_n}{q_{n+1}}\right)t^{-1/2} + O(t^{1/2})\right\}$$

Now $\quad \dfrac{p_n}{q_{n+1}} = \lim\limits_{\omega \to \infty} i\omega J(i\omega) = \lim\limits_{\omega \to \infty} \omega J_2$

and $\quad \dfrac{p_n}{q_{n+1}}\left(\dfrac{p_{n-1}}{p_n} - \dfrac{q_n}{q_{n+1}}\right) = -\lim\limits_{\omega \to \infty} \omega^2 J_1.$

Therefore

$$\sigma(x, t) = -\frac{I\sqrt{\rho}\sqrt{\lim\limits_{\omega \to \infty} \omega J_2}\, x}{2\sqrt{\pi}} \exp\left\{-\rho(\lim\limits_{\omega \to \infty} \omega J_2)\frac{x^2}{4t}\right\}$$
$$\times \left(t^{-3/2} + \frac{\lim\limits_{\omega \to \infty} w^2 J_1}{(\lim\limits_{\omega \to \infty} \omega J_2)^2} t^{-1/2} + O(t^{1/2})\right). \quad (25)$$

But the first term in eq. (25) is the stress distribution in a viscous material ($J_1 = 0$, $\omega J_2 = 1/\eta$). Hence for small values of the time the stress distribution in a material, not exhibiting instantaneous elasticity in tension, is the same as that for a material which is purely viscous in tension. The physical reason for this effect is that the strain tends to zero as $t \to +0$ and therefore only the dashpots and not the springs of the fundamental network are stressed initially.

Analytic solutions to this problem for the Voigt element and the two three-element models can be deduced from the paper of Lee and Morrison [4] by taking the function $\Sigma'(\xi, \tau)$ in that paper and differentiating it with respect to τ. This differentiation is necessary because the boundary condition considered in that paper was $\sigma(0, t) = H(t)$, whereas $\sigma(0, t) = \delta(t) = dH(t)/dt$ has been considered in this section.

An analytic solution to eq. (15) has been found† when $J(i\omega)$ is given by eq. (2, 112) with $\nu = 2/3$,

i.e. $$J(i\omega) = \frac{1}{K}(i\omega)^{-2/3}.$$

On substitution into eq. (15),

$$\sigma = -\frac{I}{\pi} R\left[\int_0^\infty \exp\{i\omega t - \rho^{1/2}K^{-1/2}x(i\omega)^{2/3}\}\, d\omega\right].$$

On integration

$$\left.\begin{array}{c}\sigma = -\dfrac{I}{t} F_{2/3}\left(\dfrac{\rho^{1/2}x}{K^{1/2}t^{2/3}}\right),\\[2ex]\text{where}\\[1ex]F_{2/3}(z) = 2\exp\left(-\dfrac{2z^3}{27}\right)\left[\dfrac{z^2}{3^{4/3}}Ai\left(\dfrac{z^2}{3^{4/3}}\right) - \dfrac{z}{3^{2/3}}Ai'\left(\dfrac{z^2}{3^{4/3}}\right)\right],\end{array}\right\} \quad (26)$$

$Ai(y)$ is the Airy function and $Ai'(y)$ denotes $\dfrac{dAi(y)}{dy}$.

Another method of solution to this problem has been used by Glauz and Lee [7]. It is suitable only when the stress–strain law in differential form is known explicitly and it contains but a few terms. The equations are referred to the characteristics and then integrated along them numerically. For the four element model given by eq. (1.10), it follows from eq. (10) of this chapter that the characteristics are $x \pm \sqrt{(q_2/\rho p_2)}t = \text{constant}$ and $x = \text{constant}$. For details of the method the reader is referred to the original paper.

3. Normal impact on the boundary of a spherical cavity in an infinite medium

This is a spherically symmetric problem. The governing differential equation for the radial displacement in an elastic medium is

$$\frac{\partial^2 u}{\partial r^2} + \frac{2}{r}\frac{\partial u}{\partial r} - \frac{2}{r^2}u = \frac{\rho}{\lambda + 2\mu}\frac{\partial^2 u}{\partial t^2}. \qquad (27)$$

† The solution is due to Mr E. J. Watson of Manchester University.

The normal radial stress σ_{rr}, denoted here by σ, is

$$\sigma = (\lambda + 2\mu)\frac{\partial u}{\partial r} + 2\lambda \frac{u}{r}. \tag{28}$$

Apply the one-sided Fourier transform to eq. (27):

$$\frac{d^2\bar{u}}{dr^2} + \frac{2}{r}\frac{d\bar{u}}{dr} - \left(\frac{2}{r^2} - \omega^2 a^2\right)\bar{u} = 0, \tag{29}$$

where

$$a^2 = \frac{\rho}{\lambda + 2\mu}. \tag{30}$$

The solution of eq. (29) which represents an outward spreading wave is

$$\bar{u} = A\frac{1 + i\omega ra}{r^2}\exp(-i\omega ra),$$

whence

$$\bar{\sigma} = A\left(\frac{\rho\omega^2}{r} - 4\mu\frac{1 + i\omega ra}{r^3}\right)\exp(-i\omega ra).$$

We consider the case when the impact is an impulsive pressure $I\delta(t)$ at the cavity face, radius R,

i.e. $\bar{\sigma} = -I$ when $r = R$.

Therefore

$$\bar{\sigma} = -I\left\{\frac{\rho\omega^2}{r} - 4\mu\frac{1 + i\omega ra}{r^3}\right\}\left\{\frac{\rho\omega^2}{R} - 4\mu\frac{1 + i\omega Ra}{R^3}\right\}^{-1}$$

$$\times \exp\{-i\omega(r - R)a\}$$

and† $\sigma = -\frac{1}{\pi}IR_e\Bigg[\int_0^\infty \exp(i\omega t)\left\{\frac{\rho(i\omega)^2}{r} + 4\mu\frac{1 + i\omega ra}{r^3}\right\}$

$$\times \left\{\frac{\rho(i\omega)^2}{R} + 4\mu\frac{1 + i\omega Ra}{R^3}\right\}^{-1}\exp\{-i\omega(r - R)a\}\,d\omega\Bigg] \tag{31}$$

The viscoelastic solution is obtained by replacing $\lambda + 2\mu$ and 2μ by the corresponding complex moduli. From Table (3.1) $\lambda + 2\mu$ is replaced by $\frac{1}{3}(Y_v + 2Y_s)$ and 2μ by Y_s. If

† $R_e[z]$ is used in the following two equations to denote the real part of z to avoid confusion with the radius R.

DYNAMIC PROBLEMS 107

$J^*(i\omega)$ denotes the complex compliance $3(Y_\nu + 2Y_s)^{-1}$, from eq. (30) $a = \sqrt{\{\rho J^*(i\omega)\}}$ and eq. (31) is modified to give the stress in a viscoelastic medium as

$$\sigma = -IR_e\left[\int_0^\infty \exp(i\omega t)\left[\frac{\rho(i\omega)^2}{r} + 2Y_s\frac{1+i\omega r\sqrt{\{\rho J^*(i\omega)\}}}{r^3}\right]\right.$$

$$\times \left[\frac{\rho(i\omega)^2}{R} + 2Y_s\frac{1+i\omega R\sqrt{\{\rho J^*(i\omega)\}}}{R^3}\right]^{-1}$$

$$\left.\times \exp[-i\omega\{\rho J^*(i\omega)\}^{1/2}(r-R)]\right]. \quad (32)$$

If eq. (32) is compared with eq. (15), which gives the stress caused by an impulse on the end of a viscoelastic rod, it is seen that, if $r - R$ is replaced by x and $J^*(i\omega)$ by $J(i\omega)$, the equations only differ by the presence of the two terms in braces in eq. (32).

$J^*(i\omega)$ can be expressed in the form $P^*(i\omega)/Q^*(i\omega)$. If the orders of P^* and Q^* are the same with coefficients of the highest order terms p_n^* and q_n^* respectively, then the lines

$$r - R \pm \sqrt{\left\{\frac{q_n^*}{\rho p_n^*}\right\}}t = \text{constant}$$

are the characteristics of the viscoelastic analogue of eq. (27). The front of any disturbance travels with velocity $\sqrt{(q_r^*/\rho p_r^*)}$.

The asymptotic expansion as $i\omega \to \infty$ of

$$\left[\frac{\rho(i\omega)^2}{r} + 2Y_s\frac{1+i\omega r\sqrt{\{\rho J^*(i\omega)\}}}{r^3}\right]$$

$$\times \left[\frac{\rho(i\omega)^2}{R} + 2Y_s\frac{1+i\omega R\sqrt{\{\rho J^*(i\omega)\}}}{R^3}\right]^{-1}$$

is†

$$\frac{R}{r}\left\{1 - \frac{2}{\sqrt{\rho}}\sqrt{\left(\lim_{\omega\to\infty} J^*\right)}\lim_{\omega\to\infty} Y_s\left(\frac{1}{R} - \frac{1}{r}\right)\frac{1}{i\omega} + O\left(\frac{1}{\omega^2}\right)\right\}.$$

Hence, using the asymptotic expansion of $\exp[-i\omega\sqrt{\{\rho J(i\omega)\}}x]$ developed in the last section, the asymptotic expansion of the

† It can be shown that, since $J^*(i\omega) \to$ constant as $\omega \to \infty$, $Y_s(i\omega) \to$ constant as $\omega \to \infty$.

integrand of eq. (32) is

$$\frac{R}{r} \exp\left[i\omega\{t - \alpha^*(r-R)\}\right] \exp\{-\beta^*(r-R)\}$$

$$\times \left[1 + \frac{1}{i\omega}\left\{\gamma^*(r-R) - \frac{2\alpha^*}{\rho}\lim_{\omega\to\infty} Y_s\left(\frac{1}{R} - \frac{1}{r}\right)\right\} + O\left(\frac{1}{\omega^2}\right)\right] \tag{33}$$

where, analogous to eqs. (20),

$$\alpha^* = \sqrt{\left(\rho \frac{p_n^*}{q_n^*}\right)} = \sqrt{\left(\rho \lim_{\omega\to\infty} J_1^*\right)}, \tag{34}$$

$$\beta^* = \frac{\rho}{2\alpha^*} \lim_{\omega\to\infty} \omega J_2^* \tag{35}$$

and $$\gamma^* = \frac{\rho}{2\alpha^*} \lim_{\omega\to\infty} \omega^2\left(J_1^* - \frac{p_n^*}{q_n^*}\right) + \frac{\beta^{*2}}{2\alpha^*}. \tag{36}$$

It follows from eq. (33), by analogy with the argument of the last section, that a normal impulse on a spherical cavity in a viscoelastic material which exhibits instantaneous elasticity is transmitted radially outwards with velocity $1/\alpha^*$ and amplitude proportional to $(1/r)\exp\{-\beta^*(r-R)\}$. The attenuated impulse is followed immediately by a wave of finite amplitude whose magnitude at its head is

$$\left(\gamma^* - \frac{2\alpha^*}{\rho r R} \lim_{\omega\to\infty} Y_s\right)(r-R) \exp\{-\beta^*(r-R)\}.$$

For a cavity in a material, for which $J^*(i\omega) = O(1/\omega)$ as $\omega \to \infty$,[†] results can be derived analogous to those for a rod for which $J(i\omega) = O(1/\omega)$ as $\omega \to \infty$. The solution for small times is

$$\sigma(r-R, t) = -\frac{I\sqrt{\rho \lim_{\omega\to\infty} \omega J_2^*}}{2\sqrt{\pi}} \frac{(r-R)R}{r t^{3/2}}$$

$$\times \exp\left\{-\rho\left(\lim_{\omega\to\infty} \omega J_2^*\right)\frac{(r-R)^2}{4t}\right\}. \tag{37}$$

[†] For $J^*(i\omega) = O(1/\omega)$ as $\omega \to \infty$, either one or both of $J_\nu(i\omega)$ and $J_s(i\omega)$ is $O(1/\omega)$ as $\omega \to \infty$.

4. Normal impact on a clamped circular plate

In the last two sections by using asymptotic expansions as $\omega \to \infty$ explicit solutions were obtained for small values of the time. In this section the final shape of the plate will be determined, i.e. large values of the time are to be considered and therefore the expansion of the transform for small values of ω is required.

The equation for the transverse displacement w of an elastic plate of thickness h in the case of radial symmetry is

$$\left(\frac{\partial^2}{\partial r^2} + \frac{1}{r}\frac{\partial}{\partial r}\right)^2 w + \kappa^2 \frac{\partial^2 w}{\partial t^2} = \frac{1}{D} Z(r, t), \tag{38}$$

where $Z(r, t)$ is the transverse load per unit area,

$$D = (2EL^3)/\{3(1 - \nu^2)\}, \tag{39}$$

E is Young's modulus and ν Poisson's ratio,

and
$$\kappa^2 = \frac{2\rho h}{D}. \tag{40}$$

If the plate is of radius a and subject to a paraboloidal impulsive load, then

$$Z(r, t) = k(a^2 - r^2)\delta(t), \quad k \text{ constant}; \tag{41}$$

and the boundary conditions are

$$w = 0 \quad \text{and} \quad \frac{\partial w}{\partial r} = 0 \quad \text{at} \quad r = a. \tag{42}$$

Applying the one-sided Fourier transform to eqs. (38), (41) and (42),

$$\left(\frac{d^2}{dr^2} + \frac{1}{r}\frac{d}{dr}\right)^2 \bar{w} - \omega^2 \kappa \bar{w} = \lambda(a^2 - r^2), \tag{43}$$

where
$$\lambda = \frac{k}{D} \tag{44}$$

and
$$\bar{w} = 0 \quad \text{and} \quad \frac{d\bar{w}}{dr} = 0 \quad \text{at} \quad r = a. \tag{45}$$

The most general solution of eq. (43) that is finite at $r = 0$ is†

$$\bar{w} = AJ_0\{\sqrt{(\kappa\omega)}r\} + BI_0\{\sqrt{(\kappa\omega)}r\} - \frac{\lambda}{\kappa^2\omega^2}(a^2 - r^2). \tag{46}$$

† J_0, J_1, I_0 and I_1 are Bessel functions.

The constants of integration A and B are determined by eqs. (45). Hence

$$\bar{w} = \frac{2a\lambda}{(\kappa\omega)^{5/2}} \frac{I_0\{\sqrt{(\kappa\omega)}a\}J_0\{\sqrt{(\kappa\omega)}r\} - J_0\{\sqrt{(\kappa\omega)}a\} I_0\{\sqrt{(\kappa\omega)}r\}}{J_1\{\sqrt{(\kappa\omega)}a\}I_0\{\sqrt{(\kappa\omega)}a\} + J_0\{\sqrt{(\kappa\omega)}a\}I_1\{\sqrt{(\kappa\omega)}a\}}$$

$$- \frac{\lambda}{\kappa^2\omega^2}(a^2 - r^2). \quad (47)$$

w is found by inverting the transform.

An elastic plate will oscillate indefinitely and therefore the inverse transform of the small ω approximation will not converge as $t \to \infty$. However eq. (47) can still be used for a viscoelastic plate provided the elastic constants are replaced by the corresponding viscoelastic moduli. Since a viscoelastic plate dissipates energy and since the total energy of the system is finite, equal to the kinetic energy of the impulse, the small ω approximation gives an inverse transform that will converge as $t \to \infty$ in the viscoelastic case.†

The complex modulus required is that corresponding to $E/(1-\nu^2)$. It will be denoted by $Y(i\omega)$. From Table (3.1),

$$Y(i\omega) = \frac{Y_s(2Y_\nu + Y_s)}{Y_\nu + 2Y_s}. \quad (48)$$

The expansion for small ω will depend on whether or not the material exhibits long term viscous flow.

If so, $Y(i\omega) = \dfrac{Q(i\omega)}{P(i\omega)} = \dfrac{q_1(i\omega) + q_2(i\omega)^2 + \ldots}{p_0 + p_1(i\omega) + p_2(i\omega)^2 + \ldots}$

$$= \frac{q_1}{p_0}(i\omega)\left(1 + \left(\frac{q_2}{q_1} - \frac{p_1}{p_0}\right)i\omega + O(\omega^2)\right); \quad (49)$$

if not, $Y(i\omega) = \dfrac{Q(i\omega)}{P(i\omega)} = \dfrac{q_0 + q_1(i\omega) + q_2(i\omega)^2 + \ldots}{p_0 + p_1(i\omega) + p_2(i\omega)^2 + \ldots}$

$$= \frac{q_0}{p_0}\left(1 + \left(\frac{q_1}{q_0} - \frac{p_1}{p_0}\right)i\omega + O(\omega^2)\right). \quad (50)$$

† In place of this physical argument, it could be shown that all the poles of the right hand side of equation (47) have positive imaginary parts for viscoelastic materials.

For small $\sqrt{(\kappa\omega)}a$, expansion of the Bessel functions in eq. (47) gives

$$\bar{w} = \frac{(a^2 - r^2)^2(7a^2 - r^2)\lambda}{2^6 \cdot 3^2} \{1 + O(\kappa^2 a^4 \omega^2)\} \quad (51)$$

From eqs. (44) and (39),

$$\lambda = \frac{3k}{2h^3} \{Y(i\omega)\}^{-1}.$$

Using eqs. (49) and (50),

$$\lambda = \begin{cases} \dfrac{3k}{2h^3} \dfrac{p_0}{q_1 i\omega} \left\{1 - \left(\dfrac{q_2}{q_1} - \dfrac{p_1}{p_0}\right) i\omega + O(\omega^2)\right\}, & q_0 = 0 \\ \dfrac{3k}{2h^3} \dfrac{p_0}{q_0} \left\{1 - \left(\dfrac{q_1}{q_0} - \dfrac{p_1}{p_0}\right) i\omega + O(\omega^2)\right\}, & q_0 \neq 0. \end{cases} \quad (52)$$

From eqs. (40) and (44), $\kappa^2 = (2\rho h/k)\lambda$. Hence $\kappa\omega$ is $O(\omega^{1/2})$ or $O(\omega)$ as $\omega \to 0$ and the expansion used in deriving eq. (51) is valid for small ω. Substitute for λ and κ^2 in eq. (51). Since $\lim_{\omega \to 0} i\omega \bar{x}(i\omega) = \lim_{t \to \infty} x(t)$,[†]

$$w \xrightarrow[t \to \infty]{} \frac{(a^2 - r^2)^2(7a^2 - r^2)}{384} \frac{kp_0}{q_1 h^3} \text{ if } q_0 = 0 \quad (53)$$

and $\quad w \xrightarrow[t \to \infty]{} 0 \text{ if } q_0 \neq 0.$

The result is that the final shape of the plate depends only upon whether the material does or does not exhibit long term viscous flow. If it does, the final shape is given by equation (53). If it does not, the plate returns to its original shape after the kinetic energy of the impulse has been dissipated. The constant in eq. (53), dependent on the material properties of the plate is p_0/q_1. It can be seen from eqs. (2.87) and (2.62) that $p_0/q_1 = 1/\eta$. Hence the final displacement of the plate is proportional to the magnitude of the impulse, inversely proportional to the long term viscous flow constant and inversely proportional to the cube of its thickness.

The reader will note that in deriving the final shape of the

† For a proof of this result, see e.g. [6].

plate the Laplace transform could well have been used. The reason for using the Fourier transform in this problem is that if it were desired to evaluate the displacement for other values of t, not very large, then it would be necessary to evaluate eq. (47) numerically using the measured values of the complex moduli. For very small values of t, a large ω approximation to eq. (47) would be suitable and this involves the values of the complex moduli at high frequencies—the creep function, whose Laplace transform is introduced by using the Laplace transform in this problem, cannot be measured for time intervals less than about one second. However, this objection does not apply, either in this chapter, or in the previous chapter, if the stress–strain law is known explicitly in differential form. It is then immaterial whether the one-sided Laplace or the one-sided Fourier transform is used.

Note added in proof:

Kolsky [8] has carried out experiments in which an explosive charge was detonated at one end of a viscoelastic rod. The resulting pulse travelled backwards and forwards along the rod with successive reflections at the two ends. The displacement at the far end of the rod was measured and the shape of the pulse before and after two traverses of the rod deduced. Kolsky calculated the change of shape from measured values of the complex modulus using a Fourier integral. The two sets of values were in good agreement.

REFERENCES

1. H. KOLSKY: *Stress waves in solids* (p. 60, Figure 14, Clarendon Press, Oxford, 1953).
2. E. H. LEE and I. KANTER: *J. Appl. Phys.* **24** (1953) 115.
3. J. A. MORRISON: *Quart. Appl. Math.* **14** (1956) 153.
4. E. H. LEE and J. A. MORRISON: *J. Polymer. Sci.* **19** (1956) 93.
5. E. T. WHITTAKER and G. N. WATSON: *Modern Analysis*, 1927, 4th ed. Cambridge University Press.
6. H. S. CARSLAW and J. C. JAEGER: *Operational methods in applied mathematics*, Second Edition (p. 255, Oxford University Press).
7. R. D. GLAUZ and E. H. LEE: *J. Appl. Phys.* **25** (1954) 947.
8. H. KOLSKY: *Phil. Mag.* H **1** (1956) 693.

CHAPTER 6

MODEL FITTING TO MEASURED VALUES OF COMPLEX MODULUS OR COMPLIANCE

1. Procedure

It can be seen from Chapters 3, 4 and 5 that, in general, solutions to stress analysis problems, which do not require a final numerical evaluation of an inverse transform, can only be obtained for all values of the time for the simplest viscoelastic materials. Actual materials, apart from elastic solids and viscous liquids, rarely have a stress–strain law which in differential form contains only a few terms. For actual materials therefore an exact solution will require a final numerical integration for each set of values of the spacial and time co-ordinates of interest. However, if a Fourier time analysis of the applied stress or displacement in a particular problem for a given viscoelastic material shows that the contributions outside the frequency interval $a \leqslant \omega \leqslant b$ are negligible, then the stress analysis results for this problem will be approximately equal to those for the same problem for any other viscoelastic material that has approximately equal complex moduli in the interval (a, b). The fact that the moduli may be widely different outside the interval is of no significance because the corresponding contributions to the inverse transforms are both negligible. This assumes that both sets of moduli are reasonably well behaved outside the interval, i.e. there is no sudden increase or decrease in any modulus sufficiently significant that, even when multiplied by the small transform of the applied stress or displacement, it still contributes appreciably to an inverse transform. It has been shown in Chapter 2, Section 7, that neither moduli has pole or zero on the real ω axis.

The choice of complex moduli to approximate those of the actual material depends upon the particular boundary value

TABLE

Complex moduli and compliances of the

Name	Model(s)	Complex modulus $Y(i\omega) = Y_1 + iY_2$	
		Y_1	Y_2/ω
Elastic	E (spring)	E	0
Viscous	η (dashpot)	0	η
Maxwell	E—η (spring and dashpot in series)	$\dfrac{\omega^2\eta^2 E}{E^2 + \omega^2\eta^2}$	$\dfrac{\eta E^2}{E^2 + \omega^2\eta^2}$
Voigt	E ∥ η	E	η
First type of four-element model	(a) E_1—η_2—(E_3 ∥ η_3)	—	—
	(b) (E_3'—η_3') ∥ (E_4'—η_4')	$E_3' + E_4'$ $-\dfrac{E_3'^3}{E_3'^2 + \omega^2\eta_3'^2}$ $-\dfrac{E_4'^3}{E_4'^2 + \omega^2\eta_4'^2}$	$\dfrac{\eta_3' E_3'^2}{E_3'^2 + \omega^2\eta_3'^2}$ $+\dfrac{\eta_4' E_4'^2}{E_4'^2 + \omega^2\eta_4'^2}$
Second type of four-element model	(c) E_1 ∥ η_2 ∥ (E_3—η_3)	$E_1 + E_3$ $-\dfrac{E_3^3}{E_3^2 + \omega^2\eta_3^2}$	$\eta_2 + \dfrac{\eta_3 E_3^2}{E_3^2 + \omega^2\eta_3^2}$
	or (d) (E_3' ∥ η_3')—(E_4' ∥ η_4')	—	—
Three-element elastic	given by (a) as $\eta_2 \to \infty$, by (b) as $\eta_4' \to \infty$, by (c) as $\eta_2 \to 0$ and by (d) as $\eta_4' \to 0$.		
Three-element viscous	given by (a) as $E_1 \to \infty$, by (b) as $E_4' \to \infty$, by (c) as $E_1 \to 0$ and by (d) as $E_4' \to 0$.		

6.1.

eight simplest viscoelastic materials

Complex compliance $J(i\omega) = J_1 - iJ_2$		Associated equations
J_1	ωJ_2	
$\dfrac{1}{E}$	0	—
0	$\dfrac{1}{\eta}$	—
$\dfrac{1}{E}$	$\dfrac{1}{\eta}$	—
$\dfrac{E}{E^2 + \omega^2\eta^2}$	$\dfrac{\omega^2\eta}{E^2 + \omega^2\eta^2}$	—
$\dfrac{1}{E_1} + \dfrac{E_3}{E_3^2 + \omega^2\eta_3^2}$	$\dfrac{1}{\eta_2} + \dfrac{1}{\eta_3} - \dfrac{E_3^2/\eta_3}{E_3^2 + \omega^2\eta_3^2}$	$\dfrac{E_3}{\eta_3}J_1 + \omega J_2 = \dfrac{E_3}{E_1\eta_3} + \dfrac{1}{\eta_2} + \dfrac{1}{\eta_3}$ (1)
—	—	—
—	—	$Y_1 + \dfrac{E_3}{\eta_3}\dfrac{Y_2}{\omega} = E_1 + E_3 + \dfrac{E_3\eta_2}{\eta_3}$ (2)
$\dfrac{E_3'}{E_3'^2 + \omega^2\eta_3'^2} + \dfrac{E_4'}{E_4'^2 + \omega^2\eta_4'^2}$	$\dfrac{1}{\eta_3'} + \dfrac{1}{\eta_4'} - \dfrac{E_3'^2/\eta_3'}{E_3'^2 + \omega^2\eta_3'^2} - \dfrac{E_4'^2/\eta_4'}{E_4'^2 + \omega^2\eta_4'^2}$	—

problem considered. Generally the best choice is the one amongst all possible choices that enable explicit solutions for the problem to be found, which best fits the moduli of the actual material in the interval (a, b). It is generally true that this choice is a simple model consisting of at most four elements. It is quite possible, however, that in certain circumstances a choice of complex moduli, such as those given by eq. (2.111), would enable an explicit solution to be found and that the constants K and ν in that equation could be chosen to give a reasonable fit in the interval (a, b). It is impossible in discussing model fitting to lay down rules valid in all cases, it is only possible to state guiding principles.

H. Kolsky and Y. Y. Shi [1] give an example of how the choice of complex moduli depends upon the frequency interval (a, b). Both E. Volterra [2] and Kolsky [3] carried out measurements on similar specimens of polyethylene but the frequency range in the former case was around 10^2 sec^{-1}, whereas in the latter, it was around 3×10^4 sec^{-1}. Both found that three-element elastic models fitted their results. But in the notation of eq. (1,8) the former found η_2'/E_2' equal to $1 \cdot 5 \times 10^{-3}$ sec, whereas the latter found η_2'/E_2' equal to 2×10^{-6} sec. These results emphasize the importance of fitting models over the frequency range determined by the Fourier time analysis of the externally applied forces and displacements.

The real and imaginary parts of the complex moduli and compliances of the eight simplest viscoelastic materials are given in Table 6.1. From this table rules for fitting models to experimentally determined complex moduli or compliances in a particular frequency interval can be formulated. For an actual material a rule will be satisfied exactly only very rarely. The degree to which the model represents the material properties in any interval will depend upon the accuracy of fit. Rules 5 to 7 are applied only if none of rules 1 to 4 are satisfied. Rule 8 is applied only if none of rules 1 to 7 are satisfied.

1. If Y_1 and/or J_1 is constant in (a, b) and Y_2 and/or J_2 is zero, then the material acts as an elastic solid in (a, b).

2. If Y_1 and/or J_1 is zero in (a, b) and Y_2/ω and/or ωJ_2 are constant, then the material acts as a viscous liquid in (a, b).

3. If J_1 and ωJ_2 are constant but non-zero in (a, b), then the material acts as a Maxwell material in (a, b).

4. If Y_1 and Y_2/ω are constant but non-zero in (a, b), then the material acts as a Voigt material in (a, b).

5. If a linear relationship exists between J_1 and ωJ_2 in (a, b), but not between Y_1 and Y_2/ω, then the material acts as a four-element model of the first type in (a, b).

6. If a linear relationship exists between Y_1 and Y_2/ω in (a, b), but not between J_1 and ωJ_2, then the material acts as a four-element model of the second type in (a, b).

7. If a linear relationship exists both between J_1 and ωJ_2 and between Y_1 and Y_2/ω in (a, b), then the material acts either as a three-element elastic or as a three-element viscous model in (a, b).

8. If Y_2/Y_1 is constant, equal to $\tan \nu\pi/2$, and if $\log Y_1$ equals $\nu \log \omega$ plus another constant C in (a, b), then the material acts like a material with stress–strain law, eq. (2.107), in (a, b).

The determination of the numerical values of the elastic and viscous constants and hence of the coefficients in the differential stress–strain law when any of rules 1 to 4 are satisfied is obvious and needs no further discussion. When either rule 5 or rule 7 is satisfied, eq. (1) in Table 6.1 determines E_3/η_3 and

$$\frac{E_3}{E_1\eta_3} + \frac{1}{\eta_2} + \frac{1}{\eta_3}.$$

Since, in these circumstances

$$J_1 = \frac{1}{E_1} + \frac{1}{E_3}\left\{1 + \left(\frac{\eta_3}{E_3}\omega\right)^2\right\}^{-1}, \qquad (3)$$

a plot of J_1 against $[1 + \{(\eta_3/E_3)\omega\}^2]^{-1}$ will determine $1/E_1$ and $1/E_3$. Hence E_1, η_2, E_3 and η_3 are found. When rule 7 is satisfied, either $1/E_1 = 0$ (three-element viscous) or $1/\eta_2 = 0$ (three-element elastic). Similarly, when either rule 6 or rule 7 is satisfied, eq. (6.2) determines E_3/η_3 and $E_1 + E_3 + (E_3\eta_2)/\eta_3$ and, since in these circumstances

$$Y_1 = E_1 + E_3 - E_3\left\{1 + \left(\frac{\eta_3}{E_3}\omega\right)^2\right\}^{-1}, \qquad (4)$$

a plot of Y_1 against $[1 + \{(\eta_3/E_3)\omega\}^2]^{-1}$ will determine $E_1 + E_3$ and E_3. When rule 8 is satisfied, it follows from eq. (2.111), which can be written in the form

$$\log Y_1 = \log\left(K \cos \frac{\nu\pi}{2}\right) + \nu \log \omega,$$

that $K = \sec(\nu\pi/2)$ antilog C.

2. First example

Bland and Lee [4] attempted to fit the first type of four-element model to the measurements of the shear compliance of

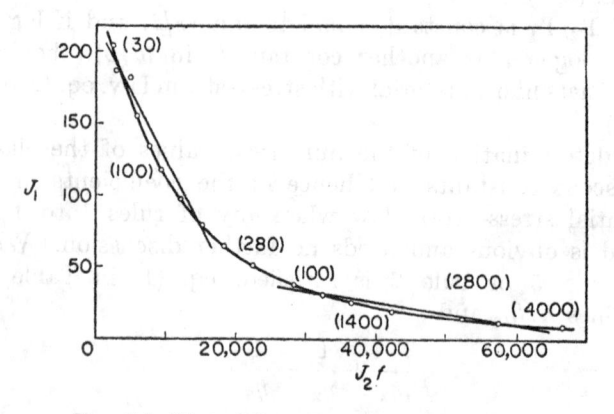

FIG. 6.1. Plot of J_1 against $J_2 f$ for polyisobutylene in shear.

polyisobutylene at 25 °C, made by Fitzgerald *et al.* [5], over a range from 30 to 4000 c/s. A plot of J_1 against $fJ_2 (\omega = 2\pi f)$ is shown in Fig. 6.1.

The plot is not sufficiently near linear to enable one model to be fitted over the entire range. However, if one decade of "fit" is all that is required, then the first type of four-element model gives a fair representation of the material properties. The straight lines corresponding to the linear J_1, ωJ_2 relationship of two such models are drawn in Fig. 6.1. From these lines the two sets of values of E_3/η_3 and $(E_3/E_1\eta_3) + 1/\eta_2 + 1/\eta_3$

are found. J_1 is plotted against the respective values of $\left\{1 + \left(\dfrac{\eta_3}{E_3}\omega\right)^2\right\}^{-1}$ in Fig. 6.2 and the values of the constants determined. The constants are given in Table 6.2.

TABLE 6.2.

Values of E_1, E_3, η_2 and η_3 for polyisobutylene at 25 °C, c.g.s. units

Frequency range	E_1	E_3	η_2	η_3
30–300 c/s	$2 \cdot 8 \times 10^7$	$6 \cdot 1 \times 10^6$	$7 \cdot 6 \times 10^4$	$8 \cdot 7 \times 10^3$
600–4000 c/s	$2 \cdot 6 \times 10^8$	$2 \cdot 4 \times 10^7$	$1 \cdot 1 \times 10^4$	$3 \cdot 0 \times 10^3$

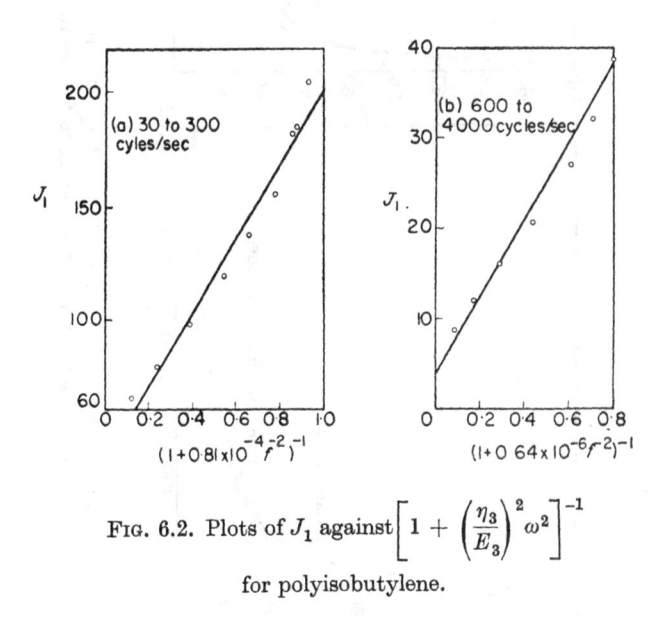

FIG. 6.2. Plots of J_1 against $\left[1 + \left(\dfrac{\eta_3}{E_3}\right)^2 \omega^2\right]^{-1}$ for polyisobutylene.

Finally J_1 and J_2 were computed from the model for the higher frequency range and compared in Fig. 6.3 with the original values.

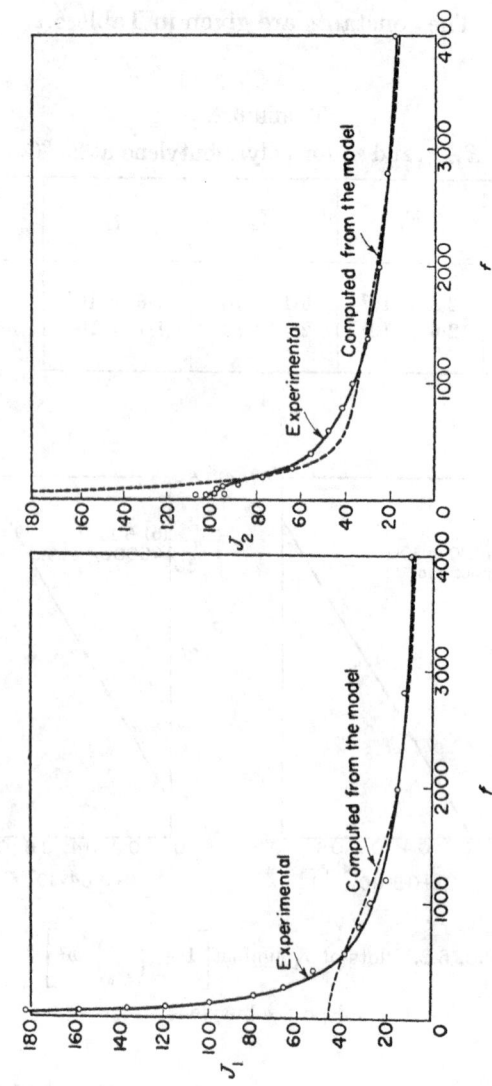

FIG. 6.3. Comparison of variation of J_1 and J_2 with frequency for polyisobutylene, measured experimentally and corresponding to four-element model of the first type.

3. Second example

Measurements in shear of Y_1 and $\tan \delta$ ($\delta = Y_2/Y_1$) for ebonite were made by Lethersich [6] and are given in Table 6.3.

Since the variation of $\tan \delta$ over seven decades of frequency is small, from rule 8 we try to fit a complex modulus of the form

$$Y(i\omega) = K(i\omega)^\nu, \quad K > 0, \quad 0 < \nu < 1. \tag{5}$$

TABLE 6.3.

Measured values of Y_1 and $\tan \delta$ for ebonite

Log frequency in c/s	$Y_1 \times 10^{-9}$, dyn/cm^2	$\tan \delta \times 10^2$	$\log Y_1$
2·788	10·1	3·1	10·004
2·334	9·9	1·6	9·996
1·544	9·3	2·3	9·968
0·000	8·9	1·3	9·949
$\bar{1}$·518	8·8	1·3	9·944
$\bar{2}$·826	8·5	1·2	9·929
$\bar{3}$·342	8·4	1·0	9·924
$\bar{4}$·518	8·2	1·7	9·914
$\bar{4}$·000	8·1	2·0	9·908

Since $\tan \delta \ll 1$, $Y_1 \gg Y_2$ and one would expect the percentage error of measurements of Y_1 to be less than those of Y_2. Therefore in Fig. 6.4 we plot $\log Y_1$ against $\log f$ ($\omega = 2\pi f$) and verify that the curve is approximately linear and that its slope is consistent with the measured values of $\tan \delta$. From (5),

$$Y_1 = K \omega^\nu \cos \frac{\nu \pi}{2}$$

and $$\log Y_1 = \nu \log \omega + \log \left(K \cos \frac{\nu \pi}{2} \right).$$

Since $\delta \ll 1$ and $\nu = (2/\pi)\delta$,

$$\nu \ll 1$$

and

$$\log Y_1 \simeq \nu \log \omega + \log K \simeq \nu \log f + \nu \log 2\pi + \log K. \tag{6}$$

From Fig. 6.4, $\nu = 0 \cdot 014$ and $\log K = 9 \cdot 957 - 0 \cdot 011 = 9 \cdot 946$, $K = 8 \cdot 83 \times 10^9$ dyn/cm^2. The fitted complex modulus is

$$Y(i\omega) = 8 \cdot 83 \times 10^9 (i\omega)^{0 \cdot 014} \text{ dyn/cm}^2. \tag{6}$$

$\tan \delta = 0 \cdot 021$, which falls within the range of the measured values in Table 6.3.

Lethersich also measured the shear strain in a creep test on ebonite under a constant stress of $7 \cdot 74 \times 10^7$ dyn/cm^2. If eq. (6.7) exactly represented the properties of the material

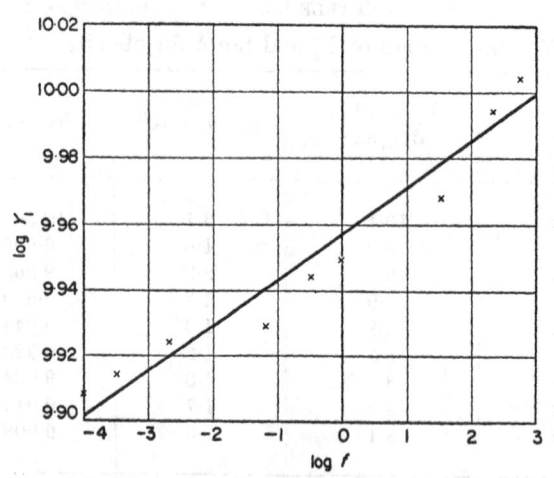

FIG. 6.4. Plot of $\log Y_1$ against $\log f$ for ebonite in shear, experimental values.

over the entire frequency range, then the creep response is given by eq. (2.105) as

$$\epsilon(t) = \frac{7 \cdot 74 \times 10^7}{8 \cdot 83 \times 10^9 \cdot \Gamma(1 \cdot 014)} t^{0 \cdot 014} H(t).$$

Now

$$t^{0 \cdot 014} = \exp(0 \cdot 014 \ln t) = \exp(0 \cdot 032 \log t)$$
$$= 1 + 0 \cdot 032 \log t + 0(\nu \ln t)^2.$$

Therefore

$$\epsilon(t) = 0 \cdot 00884 + 0 \cdot 00028 \log t + 0(\nu \ln t)^2.$$

Lethersich found that $\epsilon(t)$ is effectively linear with respect to $\log t$ for $-3 \leqslant \log t \leqslant 3$—see Fig. 3 of his paper. The straight line drawn in this figure has equation

$$\epsilon(t) = 0 \cdot 00882 + 0 \cdot 00021 \log t.$$

Hence eq. (6.7) accurately represents the magnitude of the shear strain at $t = 1$ but gives an error of about a third in its gradient with respect to log t. The gradient is small and the magnitude of $\epsilon(t)$ is predicted correct to $\pm 3\%$ over the range $-3 \leqslant \log t \leqslant 3$.

The complex modulus was fitted for periods P in the range $-3 \leqslant \log P \leqslant 4$ with an accuracy in log Y_1 of ± 0.01, i.e. with a possible fractional error of antilog $0.01 - 1$ or 2.3%.

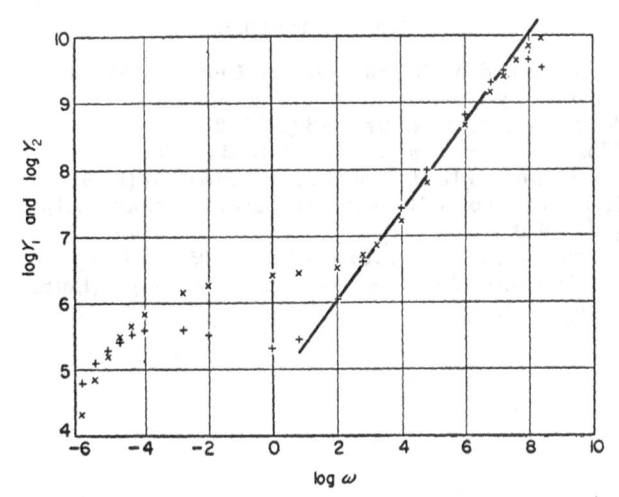

FIG. 6.5. Plot of measured values of log Y_1 and log Y_2 against log ω for polyisobutylene in shear and straight line 'fit' for $3 \leq \log \omega \leq 7$; $\times \times$ log Y_1, $++$ log Y_2.

This example shows that, if the complex modulus is fitted over the range of characteristic times required in its application, then the error introduced into the application by using the fitted complex modulus is of the same order of magnitude as the error of fitting.

That it must be confirmed that both conditions of rule 8 are satisfied before assuming that the complex modulus can be fitted by an equation of the form, $Y(i\omega) = K(i\omega)^\nu$, can be illustrated from the values of Y_1 and Y_2 for polyisobutylene in shear given by Marvin [7]. Log Y_1 and log Y_2 are plotted against log ω in Fig. 6.5. Suppose we try to fit the given equation to these values over the four decades $3 \leqslant \log \omega \leqslant 7$.

Over this range Y_1 is nearly equal to Y_2, therefore tan δ is approximately unity, $\delta \simeq \pi/4$ and $\nu = (2\delta/\pi) \simeq \frac{1}{2}$. If the equation for $Y_s(i\omega)$ had the desired form, the slope of the 'best' straight line of log Y_1 against log ω is equal to ν, i.e. approximately 0·5. But the measured value of the slope of the straight line drawn in Fig. 6.5 is approximately 0·7 and therefore an equation of the form $Y_s(i\omega) = K(i\omega)^\nu$ cannot be fitted over the four decades.

REFERENCES

1. H. KOLSKY and Y. Y. SHI: *Brown University Technical Report*, No. 5, January 1958.
2. E. VOLTERRA: *J. Appl. Mech.* **18** (1951) 273.
3. H. KOLSKY: *Proc. Phys. Soc. Lond.* B **62** (1949) 676.
4. D. R. BLAND and E. H. LEE: *J. Appl. Mech.* **23** (1956) 416.
5. E. R. FITZGERALD, L. D. GRANDINE and J. D. FERRY: *J. Appl. Phys.* **24** (1953) 650.
6. W. LETHERSICH: *Brit. J. Appl. Phys.* **1** (1950) 294.
7. R. S. MARVIN: *Proc. 2nd Int. Congr. Rheology* (Butterworths, London, 1954).

AUTHOR INDEX

Alfrey, T., 18
Andrews, R. D., 18
Atkinson, E. B., 18
Berry, D. S., 75
Biot, M. A., 18, 94
Bland, D. R., 18, 56, 75, 124
Bragg, W. L., 56
Brillouin, L., 75
Carslaw, H. S., 112
Ewing, M., 75
Ferry, J. D., 18, 75, 124
Fitzgerald, E. R., 124
Glauz, R. D., 112
Goodier, J. N., 94
Grandine, L. D., 124
Gross, B., 18, 56
Gutenberg, B., 75
Jaeger, J. C., 112
Jeffreys, H., 56
Kanter, I., 112

Kolsky, H., 75, 112, 124
Leaderman, H., 18
Lee, E. H., 75, 94, 112, 124
Lethersich, W., 124
Lomar, M., 56
Marvin, R. S., 124
Morrison, J. A., 112
Press, F., 75
Radok, J. R. M., 94
Schwarzl, L., 18
Shi, Y. Y., 124
Staverman, A. J., 18
Timoshenko, S., 94
Tobolsky, A. V., 18
Volterra, E., 124
Watson, G. N., 112
Whittaker, E. T., 112
Widder, D. V., 56
Woodward, W. B., 94

A CATALOG OF SELECTED
DOVER BOOKS
IN SCIENCE AND MATHEMATICS

Astronomy

CHARIOTS FOR APOLLO: The NASA History of Manned Lunar Spacecraft to 1969, Courtney G. Brooks, James M. Grimwood, and Loyd S. Swenson, Jr. This illustrated history by a trio of experts is the definitive reference on the Apollo spacecraft and lunar modules. It traces the vehicles' design, development, and operation in space. More than 100 photographs and illustrations. 576pp. 6 3/4 x 9 1/4. 0-486-46756-2

EXPLORING THE MOON THROUGH BINOCULARS AND SMALL TELESCOPES, Ernest H. Cherrington, Jr. Informative, profusely illustrated guide to locating and identifying craters, rills, seas, mountains, other lunar features. Newly revised and updated with special section of new photos. Over 100 photos and diagrams. 240pp. 8 1/4 x 11. 0-486-24491-1

WHERE NO MAN HAS GONE BEFORE: A History of NASA's Apollo Lunar Expeditions, William David Compton. Introduction by Paul Dickson. This official NASA history traces behind-the-scenes conflicts and cooperation between scientists and engineers. The first half concerns preparations for the Moon landings, and the second half documents the flights that followed Apollo 11. 1989 edition. 432pp. 7 x 10.
0-486-47888-2

APOLLO EXPEDITIONS TO THE MOON: The NASA History, Edited by Edgar M. Cortright. Official NASA publication marks the 40th anniversary of the first lunar landing and features essays by project participants recalling engineering and administrative challenges. Accessible, jargon-free accounts, highlighted by numerous illustrations. 336pp. 8 3/8 x 10 7/8. 0-486-47175-6

ON MARS: Exploration of the Red Planet, 1958-1978--The NASA History, Edward Clinton Ezell and Linda Neuman Ezell. NASA's official history chronicles the start of our explorations of our planetary neighbor. It recounts cooperation among government, industry, and academia, and it features dozens of photos from Viking cameras. 560pp. 6 3/4 x 9 1/4. 0-486-46757-0

ARISTARCHUS OF SAMOS: The Ancient Copernicus, Sir Thomas Heath. Heath's history of astronomy ranges from Homer and Hesiod to Aristarchus and includes quotes from numerous thinkers, compilers, and scholasticsts from Thales and Anaximander through Pythagoras, Plato, Aristotle, and Heraclides. 34 figures. 448pp. 5 3/8 x 8 1/2.
0-486-43886-4

AN INTRODUCTION TO CELESTIAL MECHANICS, Forest Ray Moulton. Classic text still unsurpassed in presentation of fundamental principles. Covers rectilinear motion, central forces, problems of two and three bodies, much more. Includes over 200 problems, some with answers. 437pp. 5 3/8 x 8 1/2. 0-486-64687-4

BEYOND THE ATMOSPHERE: Early Years of Space Science, Homer E. Newell. This exciting survey is the work of a top NASA administrator who chronicles the technological advances, the relationship of space science to general science, and the space program's social, political, and economic contexts. 528pp. 6 3/4 x 9 1/4.

0-486-47464-X

STAR LORE: Myths, Legends, and Facts, William Tyler Olcott. Captivating retellings of the origins and histories of ancient star groups include Pegasus, Ursa Major, Pleiades, signs of the zodiac, and other constellations. "Classic." – *Sky & Telescope*. 58 illustrations. 544pp. 5 3/8 x 8 1/2. 0-486-43581-4

A COMPLETE MANUAL OF AMATEUR ASTRONOMY: Tools and Techniques for Astronomical Observations, P. Clay Sherrod with Thomas L. Koed. Concise, highly readable book discusses the selection, set-up, and maintenance of a telescope; amateur studies of the sun; lunar topography and occultations; and more. 124 figures. 26 halftones. 37 tables. 335pp. 6 1/2 x 9 1/4. 0-486-42820-6

Browse over 9,000 books at www.doverpublications.com

Chemistry

MOLECULAR COLLISION THEORY, M. S. Child. This high-level monograph offers an analytical treatment of classical scattering by a central force, quantum scattering by a central force, elastic scattering phase shifts, and semi-classical elastic scattering. 1974 edition. 310pp. 5 3/8 x 8 1/2. 0-486-69437-2

HANDBOOK OF COMPUTATIONAL QUANTUM CHEMISTRY, David B. Cook. This comprehensive text provides upper-level undergraduates and graduate students with an accessible introduction to the implementation of quantum ideas in molecular modeling, exploring practical applications alongside theoretical explanations. 1998 edition. 832pp. 5 3/8 x 8 1/2. 0-486-44307-8

RADIOACTIVE SUBSTANCES, Marie Curie. The celebrated scientist's thesis, which directly preceded her 1903 Nobel Prize, discusses establishing atomic character of radioactivity; extraction from pitchblende of polonium and radium; isolation of pure radium chloride; more. 96pp. 5 3/8 x 8 1/2. 0-486-42550-9

CHEMICAL MAGIC, Leonard A. Ford. Classic guide provides intriguing entertainment while elucidating sound scientific principles, with more than 100 unusual stunts: cold fire, dust explosions, a nylon rope trick, a disappearing beaker, much more. 128pp. 5 3/8 x 8 1/2. 0-486-67628-5

ALCHEMY, E. J. Holmyard. Classic study by noted authority covers 2,000 years of alchemical history: religious, mystical overtones; apparatus; signs, symbols, and secret terms; advent of scientific method, much more. Illustrated. 320pp. 5 3/8 x 8 1/2.
0-486-26298-7

CHEMICAL KINETICS AND REACTION DYNAMICS, Paul L. Houston. This text teaches the principles underlying modern chemical kinetics in a clear, direct fashion, using several examples to enhance basic understanding. Solutions to selected problems. 2001 edition. 352pp. 8 3/8 x 11. 0-486-45334-0

PROBLEMS AND SOLUTIONS IN QUANTUM CHEMISTRY AND PHYSICS, Charles S. Johnson and Lee G. Pedersen. Unusually varied problems, with detailed solutions, cover of quantum mechanics, wave mechanics, angular momentum, molecular spectroscopy, scattering theory, more. 280 problems, plus 139 supplementary exercises. 430pp. 6 1/2 x 9 1/4. 0-486-65236-X

ELEMENTS OF CHEMISTRY, Antoine Lavoisier. Monumental classic by the founder of modern chemistry features first explicit statement of law of conservation of matter in chemical change, and more. Facsimile reprint of original (1790) Kerr translation. 539pp. 5 3/8 x 8 1/2. 0-486-64624-6

MAGNETISM AND TRANSITION METAL COMPLEXES, F. E. Mabbs and D. J. Machin. A detailed view of the calculation methods involved in the magnetic properties of transition metal complexes, this volume offers sufficient background for original work in the field. 1973 edition. 240pp. 5 3/8 x 8 1/2. 0-486-46284-6

GENERAL CHEMISTRY, Linus Pauling. Revised third edition of classic first-year text by Nobel laureate. Atomic and molecular structure, quantum mechanics, statistical mechanics, thermodynamics correlated with descriptive chemistry. Problems. 992pp. 5 3/8 x 8 1/2. 0-486-65622-5

ELECTROLYTE SOLUTIONS: Second Revised Edition, R. A. Robinson and R. H. Stokes. Classic text deals primarily with measurement, interpretation of conductance, chemical potential, and diffusion in electrolyte solutions. Detailed theoretical interpretations, plus extensive tables of thermodynamic and transport properties. 1970 edition. 590pp. 5 3/8 x 8 1/2. 0-486-42225-9

Browse over 9,000 books at www.doverpublications.com

CATALOG OF DOVER BOOKS

Engineering

FUNDAMENTALS OF ASTRODYNAMICS, Roger R. Bate, Donald D. Mueller, and Jerry E. White. Teaching text developed by U.S. Air Force Academy develops the basic two-body and n-body equations of motion; orbit determination; classical orbital elements, coordinate transformations; differential correction; more. 1971 edition. 455pp. 5 3/8 x 8 1/2. 0-486-60061-0

INTRODUCTION TO CONTINUUM MECHANICS FOR ENGINEERS: Revised Edition, Ray M. Bowen. This self-contained text introduces classical continuum models within a modern framework. Its numerous exercises illustrate the governing principles, linearizations, and other approximations that constitute classical continuum models. 2007 edition. 320pp. 6 1/8 x 9 1/4. 0-486-47460-7

ENGINEERING MECHANICS FOR STRUCTURES, Louis L. Bucciarelli. This text explores the mechanics of solids and statics as well as the strength of materials and elasticity theory. Its many design exercises encourage creative initiative and systems thinking. 2009 edition. 320pp. 6 1/8 x 9 1/4. 0-486-46855-0

FEEDBACK CONTROL THEORY, John C. Doyle, Bruce A. Francis and Allen R. Tannenbaum. This excellent introduction to feedback control system design offers a theoretical approach that captures the essential issues and can be applied to a wide range of practical problems. 1992 edition. 224pp. 6 1/2 x 9 1/4. 0-486-46933-6

THE FORCES OF MATTER, Michael Faraday. These lectures by a famous inventor offer an easy-to-understand introduction to the interactions of the universe's physical forces. Six essays explore gravitation, cohesion, chemical affinity, heat, magnetism, and electricity. 1993 edition. 96pp. 5 3/8 x 8 1/2. 0-486-47482-8

DYNAMICS, Lawrence E. Goodman and William H. Warner. Beginning engineering text introduces calculus of vectors, particle motion, dynamics of particle systems and plane rigid bodies, technical applications in plane motions, and more. Exercises and answers in every chapter. 619pp. 5 3/8 x 8 1/2. 0-486-42006-X

ADAPTIVE FILTERING PREDICTION AND CONTROL, Graham C. Goodwin and Kwai Sang Sin. This unified survey focuses on linear discrete-time systems and explores natural extensions to nonlinear systems. It emphasizes discrete-time systems, summarizing theoretical and practical aspects of a large class of adaptive algorithms. 1984 edition. 560pp. 6 1/2 x 9 1/4. 0-486-46932-8

INDUCTANCE CALCULATIONS, Frederick W. Grover. This authoritative reference enables the design of virtually every type of inductor. It features a single simple formula for each type of inductor, together with tables containing essential numerical factors. 1946 edition. 304pp. 5 3/8 x 8 1/2. 0-486-47440-2

THERMODYNAMICS: Foundations and Applications, Elias P. Gyftopoulos and Gian Paolo Beretta. Designed by two MIT professors, this authoritative text discusses basic concepts and applications in detail, emphasizing generality, definitions, and logical consistency. More than 300 solved problems cover realistic energy systems and processes. 800pp. 6 1/8 x 9 1/4. 0-486-43932-1

THE FINITE ELEMENT METHOD: Linear Static and Dynamic Finite Element Analysis, Thomas J. R. Hughes. Text for students without in-depth mathematical training, this text includes a comprehensive presentation and analysis of algorithms of time-dependent phenomena plus beam, plate, and shell theories. Solution guide available upon request. 672pp. 6 1/2 x 9 1/4. 0-486-41181-8

Browse over 9,000 books at www.doverpublications.com

CATALOG OF DOVER BOOKS

HELICOPTER THEORY, Wayne Johnson. Monumental engineering text covers vertical flight, forward flight, performance, mathematics of rotating systems, rotary wing dynamics and aerodynamics, aeroelasticity, stability and control, stall, noise, and more. 189 illustrations. 1980 edition. 1089pp. 5 5/8 x 8 1/4. 0-486-68230-7

MATHEMATICAL HANDBOOK FOR SCIENTISTS AND ENGINEERS: Definitions, Theorems, and Formulas for Reference and Review, Granino A. Korn and Theresa M. Korn. Convenient access to information from every area of mathematics: Fourier transforms, Z transforms, linear and nonlinear programming, calculus of variations, random-process theory, special functions, combinatorial analysis, game theory, much more. 1152pp. 5 3/8 x 8 1/2. 0-486-41147-8

A HEAT TRANSFER TEXTBOOK: Fourth Edition, John H. Lienhard V and John H. Lienhard IV. This introduction to heat and mass transfer for engineering students features worked examples and end-of-chapter exercises. Worked examples and end-of-chapter exercises appear throughout the book, along with well-drawn, illuminating figures. 768pp. 7 x 9 1/4. 0-486-47931-5

BASIC ELECTRICITY, U.S. Bureau of Naval Personnel. Originally a training course; best nontechnical coverage. Topics include batteries, circuits, conductors, AC and DC, inductance and capacitance, generators, motors, transformers, amplifiers, etc. Many questions with answers. 349 illustrations. 1969 edition. 448pp. 6 1/2 x 9 1/4.
0-486-20973-3

BASIC ELECTRONICS, U.S. Bureau of Naval Personnel. Clear, well-illustrated introduction to electronic equipment covers numerous essential topics: electron tubes, semiconductors, electronic power supplies, tuned circuits, amplifiers, receivers, ranging and navigation systems, computers, antennas, more. 560 illustrations. 567pp. 6 1/2 x 9 1/4. 0-486-21076-6

BASIC WING AND AIRFOIL THEORY, Alan Pope. This self-contained treatment by a pioneer in the study of wind effects covers flow functions, airfoil construction and pressure distribution, finite and monoplane wings, and many other subjects. 1951 edition. 320pp. 5 3/8 x 8 1/2. 0-486-47188-8

SYNTHETIC FUELS, Ronald F. Probstein and R. Edwin Hicks. This unified presentation examines the methods and processes for converting coal, oil, shale, tar sands, and various forms of biomass into liquid, gaseous, and clean solid fuels. 1982 edition. 512pp. 6 1/8 x 9 1/4. 0-486-44977-7

THEORY OF ELASTIC STABILITY, Stephen P. Timoshenko and James M. Gere. Written by world-renowned authorities on mechanics, this classic ranges from theoretical explanations of 2- and 3-D stress and strain to practical applications such as torsion, bending, and thermal stress. 1961 edition. 560pp. 5 3/8 x 8 1/2. 0-486-47207-8

PRINCIPLES OF DIGITAL COMMUNICATION AND CODING, Andrew J. Viterbi and Jim K. Omura. This classic by two digital communications experts is geared toward students of communications theory and to designers of channels, links, terminals, modems, or networks used to transmit and receive digital messages. 1979 edition. 576pp. 6 1/8 x 9 1/4. 0-486-46901-8

LINEAR SYSTEM THEORY: The State Space Approach, Lotfi A. Zadeh and Charles A. Desoer. Written by two pioneers in the field, this exploration of the state space approach focuses on problems of stability and control, plus connections between this approach and classical techniques. 1963 edition. 656pp. 6 1/8 x 9 1/4.
0-486-46663-9

Browse over 9,000 books at www.doverpublications.com

CATALOG OF DOVER BOOKS

Mathematics-Bestsellers

HANDBOOK OF MATHEMATICAL FUNCTIONS: with Formulas, Graphs, and Mathematical Tables, Edited by Milton Abramowitz and Irene A. Stegun. A classic resource for working with special functions, standard trig, and exponential logarithmic definitions and extensions, it features 29 sets of tables, some to as high as 20 places. 1046pp. 8 x 10 1/2. 0-486-61272-4

ABSTRACT AND CONCRETE CATEGORIES: The Joy of Cats, Jiri Adamek, Horst Herrlich, and George E. Strecker. This up-to-date introductory treatment employs category theory to explore the theory of structures. Its unique approach stresses concrete categories and presents a systematic view of factorization structures. Numerous examples. 1990 edition, updated 2004. 528pp. 6 1/8 x 9 1/4. 0-486-46934-4

MATHEMATICS: Its Content, Methods and Meaning, A. D. Aleksandrov, A. N. Kolmogorov, and M. A. Lavrent'ev. Major survey offers comprehensive, coherent discussions of analytic geometry, algebra, differential equations, calculus of variations, functions of a complex variable, prime numbers, linear and non-Euclidean geometry, topology, functional analysis, more. 1963 edition. 1120pp. 5 3/8 x 8 1/2. 0-486-40916-3

INTRODUCTION TO VECTORS AND TENSORS: Second Edition--Two Volumes Bound as One, Ray M. Bowen and C.-C. Wang. Convenient single-volume compilation of two texts offers both introduction and in-depth survey. Geared toward engineering and science students rather than mathematicians, it focuses on physics and engineering applications. 1976 edition. 560pp. 6 1/2 x 9 1/4. 0-486-46914-X

AN INTRODUCTION TO ORTHOGONAL POLYNOMIALS, Theodore S. Chihara. Concise introduction covers general elementary theory, including the representation theorem and distribution functions, continued fractions and chain sequences, the recurrence formula, special functions, and some specific systems. 1978 edition. 272pp. 5 3/8 x 8 1/2.
0-486-47929-3

ADVANCED MATHEMATICS FOR ENGINEERS AND SCIENTISTS, Paul DuChateau. This primary text and supplemental reference focuses on linear algebra, calculus, and ordinary differential equations. Additional topics include partial differential equations and approximation methods. Includes solved problems. 1992 edition. 400pp. 7 1/2 x 9 1/4. 0-486-47930-7

PARTIAL DIFFERENTIAL EQUATIONS FOR SCIENTISTS AND ENGINEERS, Stanley J. Farlow. Practical text shows how to formulate and solve partial differential equations. Coverage of diffusion-type problems, hyperbolic-type problems, elliptic-type problems, numerical and approximate methods. Solution guide available upon request. 1982 edition. 414pp. 6 1/8 x 9 1/4. 0-486-67620-X

VARIATIONAL PRINCIPLES AND FREE-BOUNDARY PROBLEMS, Avner Friedman. Advanced graduate-level text examines variational methods in partial differential equations and illustrates their applications to free-boundary problems. Features detailed statements of standard theory of elliptic and parabolic operators. 1982 edition. 720pp. 6 1/8 x 9 1/4. 0-486-47853-X

LINEAR ANALYSIS AND REPRESENTATION THEORY, Steven A. Gaal. Unified treatment covers topics from the theory of operators and operator algebras on Hilbert spaces; integration and representation theory for topological groups; and the theory of Lie algebras, Lie groups, and transform groups. 1973 edition. 704pp. 6 1/8 x 9 1/4.
0-486-47851-3

Browse over 9,000 books at www.doverpublications.com

CATALOG OF DOVER BOOKS

A SURVEY OF INDUSTRIAL MATHEMATICS, Charles R. MacCluer. Students learn how to solve problems they'll encounter in their professional lives with this concise single-volume treatment. It employs MATLAB and other strategies to explore typical industrial problems. 2000 edition. 384pp. 5 3/8 x 8 1/2. 0-486-47702-9

NUMBER SYSTEMS AND THE FOUNDATIONS OF ANALYSIS, Elliott Mendelson. Geared toward undergraduate and beginning graduate students, this study explores natural numbers, integers, rational numbers, real numbers, and complex numbers. Numerous exercises and appendixes supplement the text. 1973 edition. 368pp. 5 3/8 x 8 1/2. 0-486-45792-3

A FIRST LOOK AT NUMERICAL FUNCTIONAL ANALYSIS, W. W. Sawyer. Text by renowned educator shows how problems in numerical analysis lead to concepts of functional analysis. Topics include Banach and Hilbert spaces, contraction mappings, convergence, differentiation and integration, and Euclidean space. 1978 edition. 208pp. 5 3/8 x 8 1/2. 0-486-47882-3

FRACTALS, CHAOS, POWER LAWS: Minutes from an Infinite Paradise, Manfred Schroeder. A fascinating exploration of the connections between chaos theory, physics, biology, and mathematics, this book abounds in award-winning computer graphics, optical illusions, and games that clarify memorable insights into self-similarity. 1992 edition. 448pp. 6 1/8 x 9 1/4. 0-486-47204-3

SET THEORY AND THE CONTINUUM PROBLEM, Raymond M. Smullyan and Melvin Fitting. A lucid, elegant, and complete survey of set theory, this three-part treatment explores axiomatic set theory, the consistency of the continuum hypothesis, and forcing and independence results. 1996 edition. 336pp. 6 x 9. 0-486-47484-4

DYNAMICAL SYSTEMS, Shlomo Sternberg. A pioneer in the field of dynamical systems discusses one-dimensional dynamics, differential equations, random walks, iterated function systems, symbolic dynamics, and Markov chains. Supplementary materials include PowerPoint slides and MATLAB exercises. 2010 edition. 272pp. 6 1/8 x 9 1/4. 0-486-47705-3

ORDINARY DIFFERENTIAL EQUATIONS, Morris Tenenbaum and Harry Pollard. Skillfully organized introductory text examines origin of differential equations, then defines basic terms and outlines general solution of a differential equation. Explores integrating factors; dilution and accretion problems; Laplace Transforms; Newton's Interpolation Formulas, more. 818pp. 5 3/8 x 8 1/2. 0-486-64940-7

MATROID THEORY, D. J. A. Welsh. Text by a noted expert describes standard examples and investigation results, using elementary proofs to develop basic matroid properties before advancing to a more sophisticated treatment. Includes numerous exercises. 1976 edition. 448pp. 5 3/8 x 8 1/2. 0-486-47439-9

THE CONCEPT OF A RIEMANN SURFACE, Hermann Weyl. This classic on the general history of functions combines function theory and geometry, forming the basis of the modern approach to analysis, geometry, and topology. 1955 edition. 208pp. 5 3/8 x 8 1/2. 0-486-47004-0

THE LAPLACE TRANSFORM, David Vernon Widder. This volume focuses on the Laplace and Stieltjes transforms, offering a highly theoretical treatment. Topics include fundamental formulas, the moment problem, monotonic functions, and Tauberian theorems. 1941 edition. 416pp. 5 3/8 x 8 1/2. 0-486-47755-X

Browse over 9,000 books at www.doverpublications.com

CATALOG OF DOVER BOOKS

Mathematics-Logic and Problem Solving

PERPLEXING PUZZLES AND TANTALIZING TEASERS, Martin Gardner. Ninety-three riddles, mazes, illusions, tricky questions, word and picture puzzles, and other challenges offer hours of entertainment for youngsters. Filled with rib-tickling drawings. Solutions. 224pp. 5 3/8 x 8 1/2. 0-486-25637-5

MY BEST MATHEMATICAL AND LOGIC PUZZLES, Martin Gardner. The noted expert selects 70 of his favorite "short" puzzles. Includes The Returning Explorer, The Mutilated Chessboard, Scrambled Box Tops, and dozens more. Complete solutions included. 96pp. 5 3/8 x 8 1/2. 0-486-28152-3

THE LADY OR THE TIGER?: and Other Logic Puzzles, Raymond M. Smullyan. Created by a renowned puzzle master, these whimsically themed challenges involve paradoxes about probability, time, and change; metapuzzles; and self-referentiality. Nineteen chapters advance in difficulty from relatively simple to highly complex. 1982 edition. 240pp. 5 3/8 x 8 1/2. 0-486-47027-X

SATAN, CANTOR AND INFINITY: Mind-Boggling Puzzles, Raymond M. Smullyan. A renowned mathematician tells stories of knights and knaves in an entertaining look at the logical precepts behind infinity, probability, time, and change. Requires a strong background in mathematics. Complete solutions. 288pp. 5 3/8 x 8 1/2.

0-486-47036-9

THE RED BOOK OF MATHEMATICAL PROBLEMS, Kenneth S. Williams and Kenneth Hardy. Handy compilation of 100 practice problems, hints and solutions indispensable for students preparing for the William Lowell Putnam and other mathematical competitions. Preface to the First Edition. Sources. 1988 edition. 192pp. 5 3/8 x 8 1/2. 0-486-69415-1

KING ARTHUR IN SEARCH OF HIS DOG AND OTHER CURIOUS PUZZLES, Raymond M. Smullyan. This fanciful, original collection for readers of all ages features arithmetic puzzles, logic problems related to crime detection, and logic and arithmetic puzzles involving King Arthur and his Dogs of the Round Table. 160pp. 5 3/8 x 8 1/2.

0-486-47435-6

UNDECIDABLE THEORIES: Studies in Logic and the Foundation of Mathematics, Alfred Tarski in collaboration with Andrzej Mostowski and Raphael M. Robinson. This well-known book by the famed logician consists of three treatises: "A General Method in Proofs of Undecidability," "Undecidability and Essential Undecidability in Mathematics," and "Undecidability of the Elementary Theory of Groups." 1953 edition. 112pp. 5 3/8 x 8 1/2. 0-486-47703-7

LOGIC FOR MATHEMATICIANS, J. Barkley Rosser. Examination of essential topics and theorems assumes no background in logic. "Undoubtedly a major addition to the literature of mathematical logic." – *Bulletin of the American Mathematical Society*. 1978 edition. 592pp. 6 1/8 x 9 1/4. 0-486-46898-4

INTRODUCTION TO PROOF IN ABSTRACT MATHEMATICS, Andrew Wohlgemuth. This undergraduate text teaches students what constitutes an acceptable proof, and it develops their ability to do proofs of routine problems as well as those requiring creative insights. 1990 edition. 384pp. 6 1/2 x 9 1/4. 0-486-47854-8

FIRST COURSE IN MATHEMATICAL LOGIC, Patrick Suppes and Shirley Hill. Rigorous introduction is simple enough in presentation and context for wide range of students. Symbolizing sentences; logical inference; truth and validity; truth tables; terms, predicates, universal quantifiers; universal specification and laws of identity; more. 288pp. 5 3/8 x 8 1/2. 0-486-42259-3

Browse over 9,000 books at www.doverpublications.com

CATALOG OF DOVER BOOKS

Mathematics–Algebra and Calculus

VECTOR CALCULUS, Peter Baxandall and Hans Liebeck. This introductory text offers a rigorous, comprehensive treatment. Classical theorems of vector calculus are amply illustrated with figures, worked examples, physical applications, and exercises with hints and answers. 1986 edition. 560pp. 5 3/8 x 8 1/2. 0-486-46620-5

ADVANCED CALCULUS: An Introduction to Classical Analysis, Louis Brand. A course in analysis that focuses on the functions of a real variable, this text introduces the basic concepts in their simplest setting and illustrates its teachings with numerous examples, theorems, and proofs. 1955 edition. 592pp. 5 3/8 x 8 1/2. 0-486-44548-8

ADVANCED CALCULUS, Avner Friedman. Intended for students who have already completed a one-year course in elementary calculus, this two-part treatment advances from functions of one variable to those of several variables. Solutions. 1971 edition. 432pp. 5 3/8 x 8 1/2. 0-486-45795-8

METHODS OF MATHEMATICS APPLIED TO CALCULUS, PROBABILITY, AND STATISTICS, Richard W. Hamming. This 4-part treatment begins with algebra and analytic geometry and proceeds to an exploration of the calculus of algebraic functions and transcendental functions and applications. 1985 edition. Includes 310 figures and 18 tables. 880pp. 6 1/2 x 9 1/4. 0-486-43945-3

BASIC ALGEBRA I: Second Edition, Nathan Jacobson. A classic text and standard reference for a generation, this volume covers all undergraduate algebra topics, including groups, rings, modules, Galois theory, polynomials, linear algebra, and associative algebra. 1985 edition. 528pp. 6 1/8 x 9 1/4. 0-486-47189-6

BASIC ALGEBRA II: Second Edition, Nathan Jacobson. This classic text and standard reference comprises all subjects of a first-year graduate-level course, including in-depth coverage of groups and polynomials and extensive use of categories and functors. 1989 edition. 704pp. 6 1/8 x 9 1/4. 0-486-47187-X

CALCULUS: An Intuitive and Physical Approach (Second Edition), Morris Kline. Application-oriented introduction relates the subject as closely as possible to science with explorations of the derivative; differentiation and integration of the powers of x; theorems on differentiation, antidifferentiation; the chain rule; trigonometric functions; more. Examples. 1967 edition. 960pp. 6 1/2 x 9 1/4. 0-486-40453-6

ABSTRACT ALGEBRA AND SOLUTION BY RADICALS, John E. Maxfield and Margaret W. Maxfield. Accessible advanced undergraduate-level text starts with groups, rings, fields, and polynomials and advances to Galois theory, radicals and roots of unity, and solution by radicals. Numerous examples, illustrations, exercises, appendixes. 1971 edition. 224pp. 6 1/8 x 9 1/4. 0-486-47723-1

AN INTRODUCTION TO THE THEORY OF LINEAR SPACES, Georgi E. Shilov. Translated by Richard A. Silverman. Introductory treatment offers a clear exposition of algebra, geometry, and analysis as parts of an integrated whole rather than separate subjects. Numerous examples illustrate many different fields, and problems include hints or answers. 1961 edition. 320pp. 5 3/8 x 8 1/2. 0-486-63070-6

LINEAR ALGEBRA, Georgi E. Shilov. Covers determinants, linear spaces, systems of linear equations, linear functions of a vector argument, coordinate transformations, the canonical form of the matrix of a linear operator, bilinear and quadratic forms, and more. 387pp. 5 3/8 x 8 1/2. 0-486-63518-X

Browse over 9,000 books at www.doverpublications.com

CATALOG OF DOVER BOOKS

Mathematics-Probability and Statistics

BASIC PROBABILITY THEORY, Robert B. Ash. This text emphasizes the probabilistic way of thinking, rather than measure-theoretic concepts. Geared toward advanced undergraduates and graduate students, it features solutions to some of the problems. 1970 edition. 352pp. 5 3/8 x 8 1/2. 0-486-46628-0

PRINCIPLES OF STATISTICS, M. G. Bulmer. Concise description of classical statistics, from basic dice probabilities to modern regression analysis. Equal stress on theory and applications. Moderate difficulty; only basic calculus required. Includes problems with answers. 252pp. 5 5/8 x 8 1/4. 0-486-63760-3

OUTLINE OF BASIC STATISTICS: Dictionary and Formulas, John E. Freund and Frank J. Williams. Handy guide includes a 70-page outline of essential statistical formulas covering grouped and ungrouped data, finite populations, probability, and more, plus over 1,000 clear, concise definitions of statistical terms. 1966 edition. 208pp. 5 3/8 x 8 1/2. 0-486-47769-X

GOOD THINKING: The Foundations of Probability and Its Applications, Irving J. Good. This in-depth treatment of probability theory by a famous British statistician explores Keynesian principles and surveys such topics as Bayesian rationality, corroboration, hypothesis testing, and mathematical tools for induction and simplicity. 1983 edition. 352pp. 5 3/8 x 8 1/2. 0-486-47438-0

INTRODUCTION TO PROBABILITY THEORY WITH CONTEMPORARY APPLICATIONS, Lester L. Helms. Extensive discussions and clear examples, written in plain language, expose students to the rules and methods of probability. Exercises foster problem-solving skills, and all problems feature step-by-step solutions. 1997 edition. 368pp. 6 1/2 x 9 1/4. 0-486-47418-6

CHANCE, LUCK, AND STATISTICS, Horace C. Levinson. In simple, non-technical language, this volume explores the fundamentals governing chance and applies them to sports, government, and business. "Clear and lively ... remarkably accurate." – *Scientific Monthly.* 384pp. 5 3/8 x 8 1/2. 0-486-41997-5

FIFTY CHALLENGING PROBLEMS IN PROBABILITY WITH SOLUTIONS, Frederick Mosteller. Remarkable puzzlers, graded in difficulty, illustrate elementary and advanced aspects of probability. These problems were selected for originality, general interest, or because they demonstrate valuable techniques. Also includes detailed solutions. 88pp. 5 3/8 x 8 1/2. 0-486-65355-2

EXPERIMENTAL STATISTICS, Mary Gibbons Natrella. A handbook for those seeking engineering information and quantitative data for designing, developing, constructing, and testing equipment. Covers the planning of experiments, the analyzing of extreme-value data; and more. 1966 edition. Index. Includes 52 figures and 76 tables. 560pp. 8 3/8 x 11. 0-486-43937-2

STOCHASTIC MODELING: Analysis and Simulation, Barry L. Nelson. Coherent introduction to techniques also offers a guide to the mathematical, numerical, and simulation tools of systems analysis. Includes formulation of models, analysis, and interpretation of results. 1995 edition. 336pp. 6 1/8 x 9 1/4. 0-486-47770-3

INTRODUCTION TO BIOSTATISTICS: Second Edition, Robert R. Sokal and F. James Rohlf. Suitable for undergraduates with a minimal background in mathematics, this introduction ranges from descriptive statistics to fundamental distributions and the testing of hypotheses. Includes numerous worked-out problems and examples. 1987 edition. 384pp. 6 1/8 x 9 1/4. 0-486-46961-1

Browse over 9,000 books at www.doverpublications.com